Unusual Vintage Tractors

C.H. Wendel

Published by

krause publications

700 E. State Street • Iola, WI 54990-0001
Telephone: 715/445-2214

Please call or write for our free catalog of automotive publications. Our toll-free number to place an order or obtain a free catalog is 800-258-0929 or please use our regular business telephone 715-445-2214 for editorial comment and further information.

Library of Congress Catalog Number: 96-76700
ISBN: 0-87341-410-1
Printed in the United States of America

Contents

Introduction

Compiling this book has been very enjoyable. To put it another way, it was fun for us to dig through stacks of tractor material, looking for unique designs. Sometimes these were unique because they were so innovative, and for many others, their uniqueness came about because they were so completely unconventional. In fact, some early tractor designs were simply not practical.

Tractor design wasn't limited to professional engineers. A great many designs came from farmers and mechanics who were able to convert their ideas into reality. Some of their ideas eventually made their way into other tractor designs. Unfortunately, the matter of "who made what, and when they made it" is not always an exact science. In many instances, the true facts are clouded by time.

On the other hand, there was perhaps no other company of its time that poured as much money into tractor development as International Harvester (IH). By 1905 this huge corporation had apparently committed itself to tractor development. Considering that in 1905 the farm tractor was virtually non-existent, this was indeed a huge gamble on the part of IH. Yet, the company persisted in its efforts, and this resulted in a number of very popular tractor models. By the late 1920s IH had developed the Farmall tractor, the first commercially successful row-crop, all-purpose design. Almost immediately the word "Farmall" became a familiar term in virtually every farm household.

The author has been writing books and articles about engines and tractors for a quarter century. What started as a quest for information on a few companies became a career, not by plan or design; it simply came to pass. However, during that time, it has been interesting to observe the development of the farm tractor, especially in the period up to 1950. Since that time, engines have been improved, drivetrains are more efficient, and sound-proof, air conditioned cabs have become the rule rather than the exception. However, the basic row-crop tractor designs were essentially perfected by that time, as was the three-point hitch system. In other words, it is our position that farm tractor development from the 1950s to the present has been a series of improvements and enhancements of the basic designs perfected prior to that time.

Taking a look at the development of the Farmall tractor from International Harvester, there is ample photographic evidence that this tractor went through countless experimental models and prototypes before finally achieving success as an all-around farm tractor. The tricycle chassis permitted row-crop work, yet the same tractor could go from cultivating corn to grinding feed or to hauling grain from the field. A look at numerous tractor designs of the period illustrates a few models that had the potential to be a conventional row-crop design. They had the tricycle-type chassis, but apparently their designers hadn't grasped the concept of giving their tractor a universal row-crop design.

Tractor engines are yet another matter of interest. In the developmental years, especially prior to 1920, tractor engines took on virtually every possible configuration. At first, one and two-cylinder models were the most prevalent. A few early builders used a four-cylinder engine, and even fewer employed the six-cylinder design. While tractor builders generally were loathe to "borrow" ideas from auto and truck design, they eventually did so. For the most part, the multiple cylinder tractor engines were completely redesigned from the automotive contemporaries. Heavier crankshafts and bearings were used, along with larger water passages, and a long-stroke engine design. Kerosene was a popular fuel, particularly since it was much cheaper than gasoline. However, kerosene can be atomized, but cannot be vaporized, as can gasoline. All sorts of preheating devices were used in an attempt to provide complete atomization of the fuel, but despite the best efforts a certain amount of "crankcase dilution" occurred. Simply put, unburnt kerosene made its way past the rings, down the cylinder walls, and into the crankcase. Distillate was another version, as was "tractor fuel." Both were somewhere between gasoline and kerosene, and were marketed to some degree into the 1950s.

Already in the 1940s Minneapolis-Moline developed a tractor design to utilize LP-gas. This idea made slow headway until the early 1950s, and remained available on certain tractor models into the 1960s.

Diesel engines were another subject entirely. International Harvester pioneered the use of diesel engines in farm tractors, being the first company to offer a wheel-type tractor with a diesel engine. From its inception in the 1930s, the IH diesels remained available for certain IH tractor models from that point onward. Other companies experimented with diesel tractor engines, and a few were offering a limited number to the market by the late 1940s. Diesels made slow but steady progress into the 1950s and the 1960s, gaining wide acceptance in the latter decade. By 1970, there were very few farm tractors being built with spark-ignition engines; the diesel engine had come to dominate the field.

Undoubtedly, one could digress into various phases of farm tractor development. The great deficiency within this book is that many interesting areas of tractor development remain unexplored. On the other hand, the field is so broad, that there could never be a single title which would tell everything about every tractor. Even after twenty-five years of research into the history of the farm tractor, we still come across tractor companies heretofore unlisted in our files. Yet, if this book serves to enlighten interested readers on various aspects of tractor development, we will feel well rewarded for our efforts.

As a final note, many of the illustrations used in this book have been taken from frail and aging magazines. The original artwork is, in many cases, of poor quality. Even with a bit of darkroom manipulation, they may not be of the quality one might desire. On the other hand, many of the illustrations are the only known visual record of the tractors subjected to our scrutiny.

Should the reader find errors or omissions, or have materials that might be of use for future titles, please contact the author, in care of Krause Publications.

C.H. Wendel
December 18, 1995

Acason Farm Tractor Company, Detroit, Michigan

Research by this writer has not made a connection between this firm and the Acason Motor Truck Company. However, in 1918, the Acason tractor attachment for the Ford Model T was being advertised and sold. Since there were more Model Ts on the road than anything else, the majority of companies making tractor attachments for automobiles built their designs around the venerable Model T.

Adams Sidehill Tractor Co., The Dalles, Oregon

In the mid-1920s, E.G. Adams at The Dalles, Oregon, developed a special tractor design to conquer the steep sidehills of Oregon and other western states. This resulted in the Adams Sidehill Tractor of 1927. That year, the Adams Sidehill Tractor Company was organized at The Dalles, but aside from this illustration in a 1927 issue of the *American Thresherman*, nothing further is known of the company or the inventor.

Adaptable Tractor Co., Indianapolis, Indiana

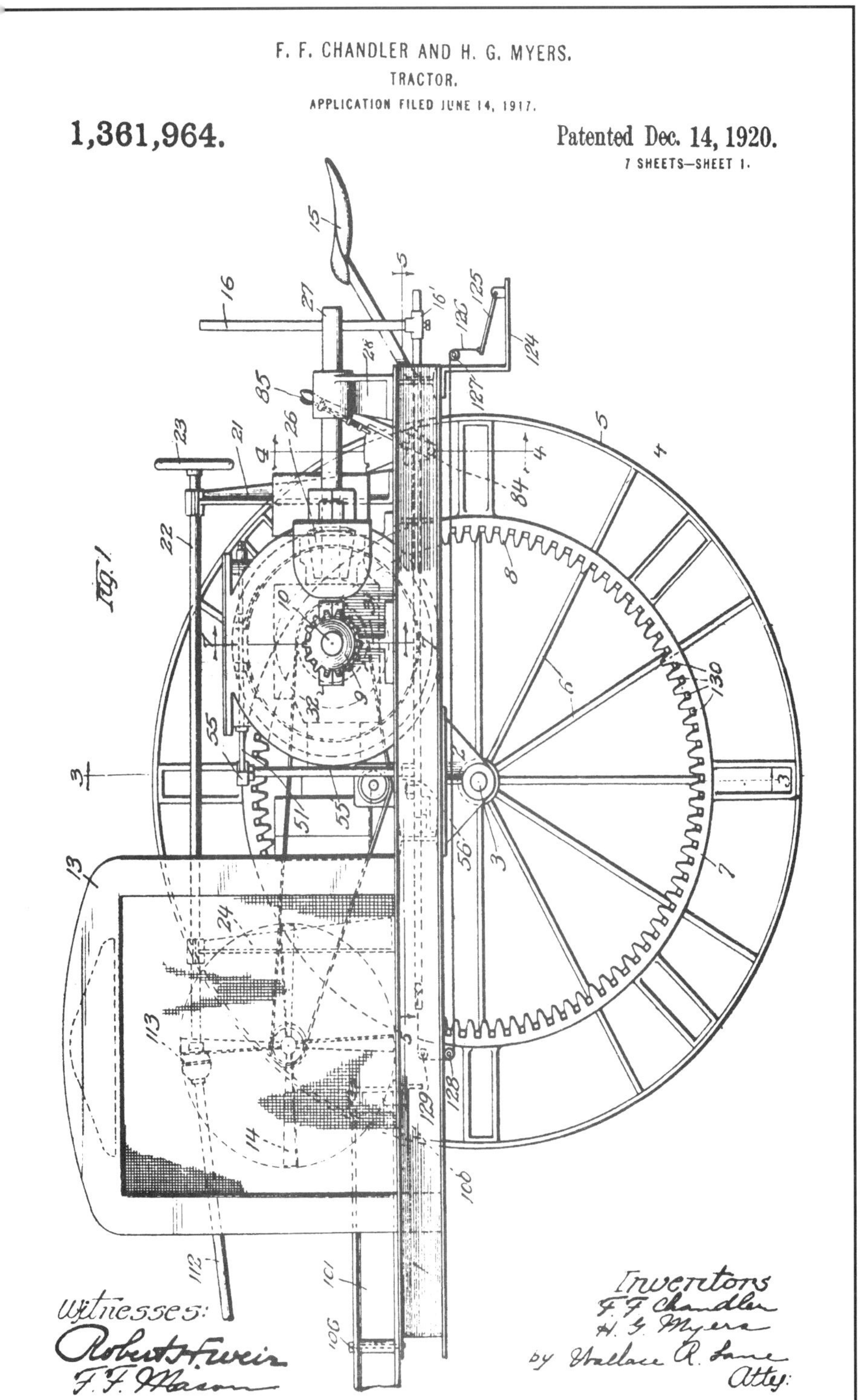

Here is a tractor patented in 1920, with the company being incorporated the same year. The Adaptable was one of many "universal" designs that emerged in the 1912-20 period. Its plan was that it be adaptable to many different farm implements as the search for the all-around general purpose tractor continued. Unfortunately, it does not appear that the company made any headway, and outside of this patent, little else is known of the Adaptable tractor.

Advance-Rumely Thresher Company, LaPorte, Indiana

The story of Advance-Rumely and their world-renowned OilPull tractor is an oft-told tale, but one that continues to captivate virtually everyone interested in the development of the farm tractor. The seeds for the OilPull were sown in the early 1900s when Dr. Edward A. Rumely was in Germany and visited with Rudolph Diesel. Rumely was determined to build tractors in America that "could burn any grade of fuel under any conditions."

Eventually, Rumely sought out John A. Secor, a famous engine designer living in New York. Secor came to LaPorte, and he, together with Wm. H. Higgins, the Rumely factory superintendent, and Dr. Rumely, cooked up the first Rumely OilPull prototype in 1908. Its success led to further examples in 1909, and the following year Advance-Rumely began marketing the OilPull in a big way. The immediate and magnificent success of the OilPull probably did more than any other single factor to toll the bell for steam traction engines.

Aside from its unique appearance, the OilPull had a sound all its own. It used oil for the coolant instead of water, and that was a plus for the farmer who had nothing but alcohol to use for antifreeze, which boiled away, requiring constant replenishment. It could successfully burn kerosene, largely because water was introduced into the cylinders to retard preignition. In fact, the OilPull would consume nearly as much water per day as kerosene.

The OilPull used a high compression engine, much higher than any of its contemporaries. The jacket temperature was somewhat higher than other tractors, and this was the major reason for the oil coolant. High compression and a high jacket temperature, together with the use of water to retard preignition, all played a role in making the OilPull one of the few farm tractors to successfully use kerosene as a fuel. Other companies tried various methods, and while some were rather successful, none could compare to the unique system embodied in the OilPull.

While the original heavyweight design was extremely popular for some years, it was obvious by the 1920s that the industry trend was toward lightweight tractors. So, in 1926 Rumely came with a series of lightweight designs. Although modified somewhat from the original style, these tractors maintained the same basic principles embodied in the original OilPull design. However, the tractor industry was changing at breakneck speed toward row-crop designs. This, coupled with the Great Depression of 1929 and the years following, led to the demise of Advance-Rumely. In 1931 the company was purchased by Allis-Chalmers and continued in operation as A-C LaPorte Works.

When this experimental Rumely tractor was built about 1908, the manufacturer was M. Rumely Company. Advance-Rumely appeared in 1915 due to a reorganization plan. This experimental model embodied many of the features that would remain with the famous OilPull throughout its production run, ending about 1930. One of the unique features of the OilPull was its ability to successfully burn kerosene fuel regardless of the load on the engine.

In 1910 the first production model of the inimitable OilPull appeared. After the first year of production it was given a rating of 25 drawbar and 45 belt horsepower. Maximum engine speed was only 375 rpm. Production of this model ended in 1914. Also during 1911 Rumely introduced its famous 30-60 OilPull of similar design. This big tractor remained in the lineup for

The early Rumely line was filled out in 1911 with the more-than 15-30 OilPull. The engine was a single cylinder design, or half of the 30-60 engine. Its 30 belt horsepower came from a 10-inch bore and 12-inch stroke, but with a crankshaft speed of only 375 rpm. One of the most unique features of the OilPull was its use of an oil coolant, rather than water. The exhaust was piped to nozzles above the cooling sections; they were cooled by the induced draft created by the engine exhaust. The system was simple, ingenious, and very effective. Numerous other OilPull models were included in the line, all of them following the same general designs.

Rumely built its OilPulls in virtually the same pattern into 1923. The smallest of these was a 12-20 model that was nearly identical in appearance, differing primarily in size. The 12-20 and 16-30 models were very popular with smaller farms, and the 20-40 was extensively used by threshermen and sawmill operators. At the top of the line, the big 30-60 was used for the largest power needs, such as big prairie plows and the largest threshing machines. Even though they haven't been built for over sixty years, the OilPull tractors still retain a certain mystique that makes them very popular at vintage tractor exhibitions.

The Advance-Rumely Thresher Company introduced its new lightweight series of the OilPull in 1924. The new tractors were noticeably different than their predecessors, but the same basic principles applied to the engine and cooling system design. Several different models were built in the period up to 1931, when Advance-Rumely was acquired by Allis-Chalmers. The latter continued to sell the OilPull until the remaining parts inventory was depleted.

In 1927 Advance-Rumely acquired the Toro tractor line in an attempt to enter the row-crop tractor business. After some redesign, the Toro was reincarnated as the Rumely DoAll tractor. This model, shown here as a cultivator, was a convertible tractor, that is, it could be converted to an ordinary standard-tread tractor if desired. A number of different implements could be used with the DoAll besides the cultivator shown here.

The Rumely DoAll tractor was one of the few convertible tractors offered to the market. This unique design could be converted back and forth as a tractor or as a cultivator. This ancient factory photograph shows the DoAll as a conventional farm tractor. While the design was quite ingenious, farmers of 1930 were already well indoctrinated to the row-crop designs then on the market. The wide acceptance of the new row-crop design made selling this convertible machine rather difficult. After the 1931 takeover by Allis-Chalmers, production of the DoAll ended, with the latter selling out the remaining inventory to close the era of the Rumely DoAll.

Allis-Chalmers Mfg. Company, Milwaukee, Wisconsin

In 1914 Allis-Chalmers was a huge corporation, but its manufacturing experience up to that time had been primarily in large industrial and manufacturing equipment. However, the firm decided to get into the tractor business, doing so with this 10-18 tractor. Sales of this model were no doubt disappointing, but the company persevered, eventually becoming one of the major players in the farm tractor industry.

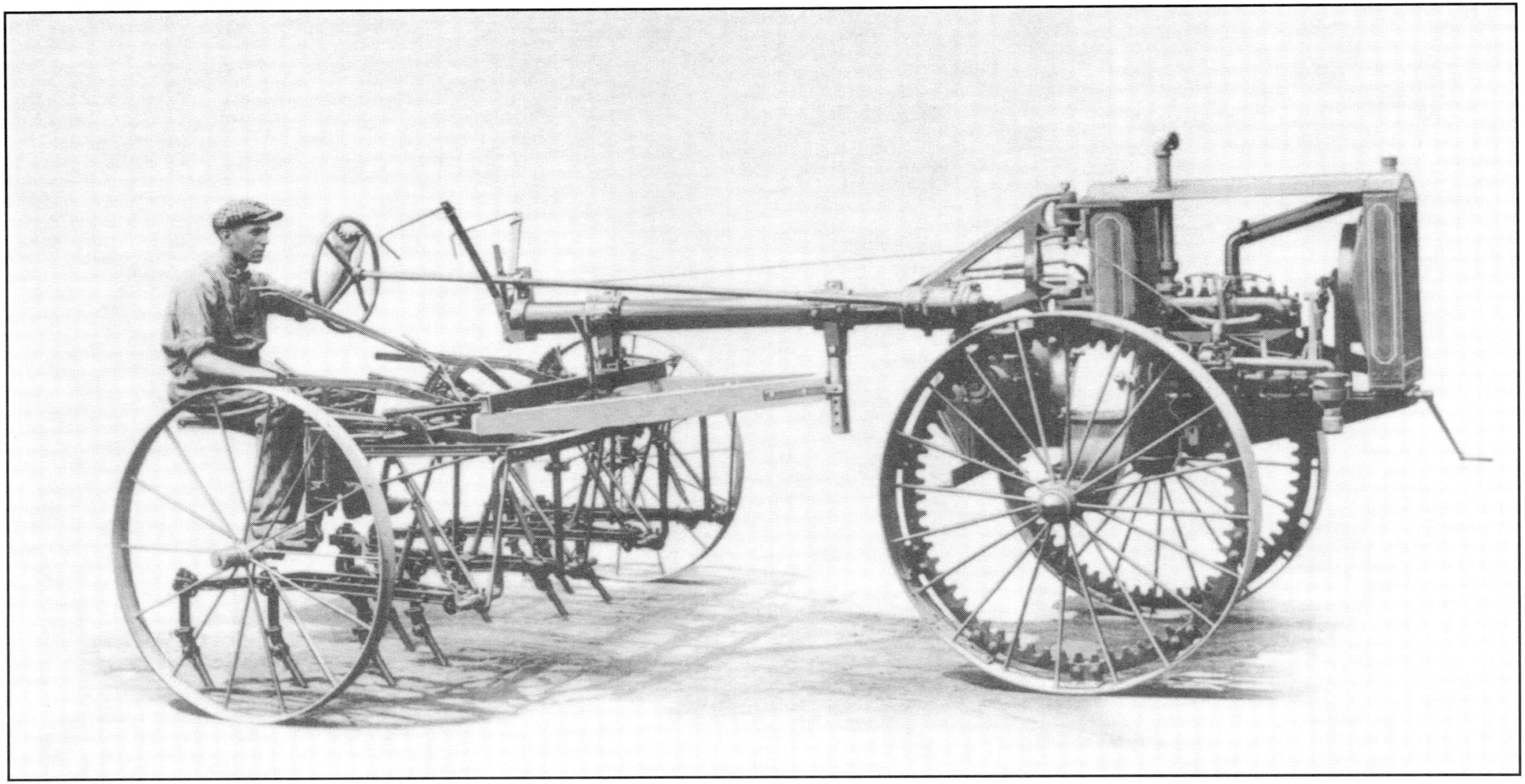

Between 1919 and 1923 Allis-Chalmers built its little 6-12 tractor. It was designed in a manner similar to several others of the period, with the drivers forward and the engine mounted between. With this system, many different implements could be adapted to the tractor. Shown here is a two-row cultivator, but plows, grain binders, wagons, and many other farm implements could be used with the 6-12.

A special low-down orchard model of the 6-12 Allis-Chalmers tractor appeared in 1921. In this design the drive wheels were mounted to special brackets in order to lower the machine. This 6-12 Model B Orchard tractor is shown with an attached plow, but presumably, other orchard equipment could also be used as needed.

By 1918 Allis-Chalmers had introduced the first of its unit frame tractor designs, continuing with various models during the 1920s. In 1929 came the famous Model U tractor; it was an entirely new design that was compact, yet very powerful for its size. The following year, 1930, saw the new Model UC, a row-crop version. From these tractor models Allis-Chalmers launched an entire series that would take the company through the 1930s.

Rubber tires were available as standard equipment on the Model U tractors, beginning in 1932. At first, much of the industry scoffed at the idea, but within a year or so there was no doubt that pneumatic tires for tractors were a revolutionary step forward. The Model U tractors were by themselves tremendously popular, but Allis-Chalmers tapped their popularity even further by demonstrating the rubber-tired versions at local fairs with staged tractor races.

Late in 1933 the Model W row-crop tractor appeared; it was soon known as the WC. This model could be equipped with a variety of mounted implements, and by the end of the 1930s Allis-Chalmers had developed a corn picker that could be mounted on the WC tractor. The WC was the first tractor in the industry to use a "square" engine, that is, the bore and stroke dimensions were the same, namely, four inches. During the 1930s Allis-Chalmers was also highly successful with its new All-Crop combine; this photograph of the late 1930s shows a WC tractor with a Model 60 All-Crop

Like other farm equipment manufacturers, Allis-Chalmers was severely hampered in regular production due to the military requirements of World War II. However, subsequent to the war, an entirely new series of tractors appeared, including the famous WD. It was followed in 1953 by the WD-45, as shown here. Allis-Chalmers pioneered the power-adjustable rear wheel system, and this alone was a major advance in tractor design.

In 1928 Allis-Chalmers purchased the Monarch Tractors Company, thereby gaining an already established and proven line of crawler tractors. The company made steady improvements in the A-C crawlers, offering an extensive line with many special equipment items available for special needs. The company's HD-15 crawler was introduced in 1951; it was the first crawler tractor available with a torque converter drive. This one is pulling a huge ripper, tearing up an old concrete roadway. In 1985 Allis-Chalmers was sold to Deutz of Germany.

Ann Arbor Hay Press Company, Ann Arbor, Michigan

With the development of the gasoline engine in the 1900-1910 period came a virtually endless list of applications for this new form of power. Ann Arbor had long been in the hay press business, and this 1910 outfit not only motorized the press mechanism, it also provided a power train to the rear wheels so that the machine could be moved from place to place under its own power. However, this unit stands as one of the early developments of self-propelled farm equipment.

Aultman & Taylor Machinery Company, Mansfield, Ohio

First offered for sale in 1911, the Aultman-Taylor was among the best tractors of its time. Although big and heavy, it was also powerful and reliable. Well over 4,000 of these 30-60 tractors were built up to 1924. This early style used an induced draft radiator whereby the engine exhaust terminated in vertical nozzles in the stack. This induced draft effect pulled cool air through the radiator for very effective cooling of the engine.

By about 1913 Aultman-Taylor abandoned the induced draft radiator, substituting a large tubular radiator and dual belt-driven fans. Except for some changes to the clutch and ignition system, the tractor remained essentially unchanged througout its entire production run. The 30-60 model shown here weighed nearly 25,000 pounds.

The Aultman-Taylor used a unique gearing system. A clutch in the belt pulley and one in the flywheel were operated by a single lever. By this method the operator could move forward or back very easily. A big four-cylinder engine was used; it had a 7 x 9 inch bore and stroke. This tractor was capable of pulling well over 9,000 pounds; its greatest disadvantage was a maximum ground speed of only 2 1/2 mph.

F.C. Austin Company, Chicago, Illinois

This curious crawler tractor design first appeared about 1919. Austin was a major manufacturer of road building machinery, and probably saw a market among its large list of contractors. The engine is mounted to the rear, with the drivetrain mounted forward. Power to the tracks was transmitted via heavy bull chains. Also of note is the peculiar design of the drive and idler sprockets.

Automotive Corporation, Toledo, Ohio

While some farmers were entirely ready to trade in their horses for a tractor, others were reluctant to give up the idea of a tractor having a steering wheel instead of the leather lines through which they controlled their horses. Thus, there were numerous companies that built a "line-drive" tractor, among them being this Automotive tractor of 1919. The two leather lines could control speed, forward and reverse, and the steering. Releasing the pressure on both lines brought the tractor to a stop.

Avery Company, Peoria, Illinois

The Avery Company had a long history, going back to 1874 and the Avery corn planter. Two brothers, Cyrus M. and Robert H. Avery, formed the Avery Planter Co. in 1874 to manufacture a corn planter of their own design, along with other small farm implements. Initially they began their work at Galesburg, Illinois, but moved to Peoria in 1884. Avery company began building steam traction engines and threshers in 1891, and within a short time these proved to be the primary products of the company.

John B. Bartholomew, a relative of the Avery family, became vice-president in 1893, and eventually came into control of the company. Bartholomew was responsible for many of the Avery engine, thresher, and tractor designs, in whole, or in part. The Avery name was widely known, as were its "Yellow Fellow" threshers, so named because they were painted bright yellow. The Avery undermounted steam engines were a unique design with the engine mounted on a chassis beneath the boiler.

The company began tractor experiments sometime prior to 1910, and the following year they began limited production. Within a short time the Avery tractors were being produced in large numbers. By 1914 the company was offering more models and sizes of tractors than virtually any of its competitors, even though this led to excessively large parts inventories.

About 1915 Avery began building various models of small tractors, some of which were of very interesting design. For reasons now unknown, Avery's rather unconventional small tractor designs never caught the market with the force of the competing Hart-Parr, or the IHC Titan, to name a couple. These disappointments, coupled with a financial panic of the early 1920s, put the firm into financial embarrassment, and eventually into bankruptcy during 1924.

After reorganizing, the company enjoyed reasonable success until 1931 and the Great Depression. Once again Avery saw new life through reorganization, continuing up to World War II.

In January 1913 Avery announced its big 40-80 tractor, meaning that it was capable of forty drawbar and eighty belt horsepower. This huge tractor had a list price of $2,650, a substantial amount of money for 1913. Many of these were used for threshing and in sawmills, while many others helped convert the Great Plains from prairie into productive farm land.

By 1919 the Avery tractor line included eight different models, including this 14-28 tractor. It was a four-cylinder model with opposed cylinders, two of them on each side of the crankshaft. Also visible is the starting lever beside the flywheel. Notches in the wheel were engaged by a pawl on the starting lever.

Tractors built by Avery had many unique features, the most obvious being the sliding frame transmission design. As shown in this old catalog illustration, the entire frame could slide forward or back to engage or disengage the transmission gearing. Also visible is the Avery four-cylinder engine of opposed cylinder design.

For forward travel, the entire engine was moved back to the intermediate gear with a long lever on the operator's platform. This is shown in the left-hand photograph. In the center, the idler gear is pulled into position between the crankshaft pinion and the intermediate gear. The right-hand photograph shows all the gearing disengaged from the crankshaft pinion; in this neutral gear the belt pulley could be engaged for threshing or other belt duties.

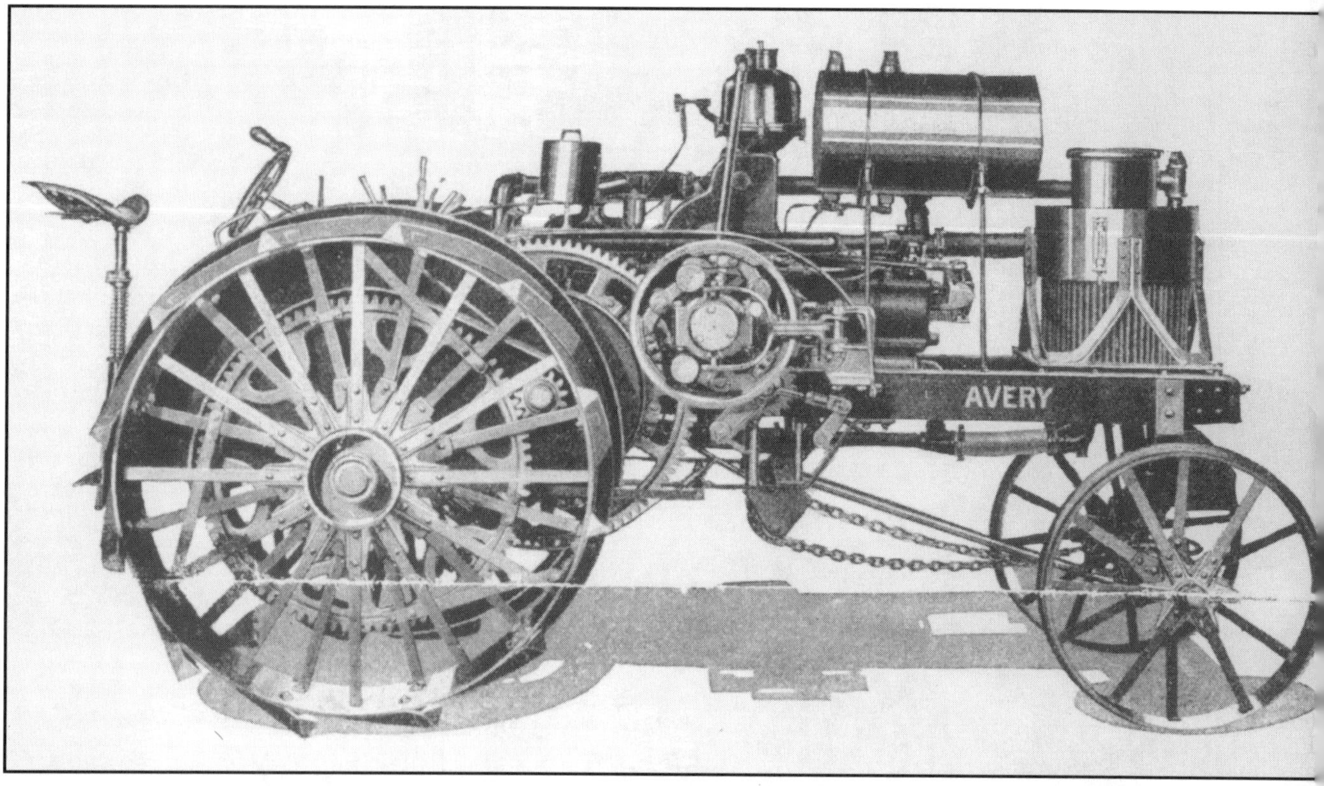

The Avery Company was early to recognize the need for a small tractor to meet the needs of the average farmer. The huge 40-80 was comparatively useless on a 160-acre farm, but this 1914 model could pull two or three plows, operate a small thresher, or a silo filler. All of the Avery tractors of this period used the same basic design and were nearly alike in outward appearance.

A 1915 Avery operator's manual notes that electrical equipment was available for "running night and day." This was a major advancement, and one that was unique for the time. However, it does not appear that this attachment gained wide popularity; the idea of working day and night was generally not approved of by the ordinary farmer of 1915.

FIG. 2

Fig. No. 2 shows the generator with the cover removed. If the commutator "E" becomes rough or dirty, smooth it by holding a piece of fine sand paper on its face while the machine is running. If the brushes, F, G, H and I, become dirty they can be lifted out and cleaned with gasoline. Use care to see that all the gasoline is removed before starting the machine.

FIG. 5

The Electric Motor, or starter, is located at the side of the engine over the flywheel. This machine is used only for the starting. It should be oiled with a good grade of machine oil about once a month, and the gear case filled with grease once or twice during the season. The metal cap at the end should be removed at the beginning of the season and the brushes and commutator cleaned of carbon and copper dust.

Electric starting and lighting equipment was available for any Avery tractor, beginning about 1915. An early operator's manual illustrates and describes these attachments, noting that "the starter is located at the side of the engine over the flywheel." Thus, the Avery stands as perhaps the only large tractor of its time to be so equipped.

In 1916 Avery announced its small 5-10 tractor. Weighing only 2,250 pounds, it was the company's first lightweight tractor design. The friction drive system is obvious directly beneath the operator's seat.

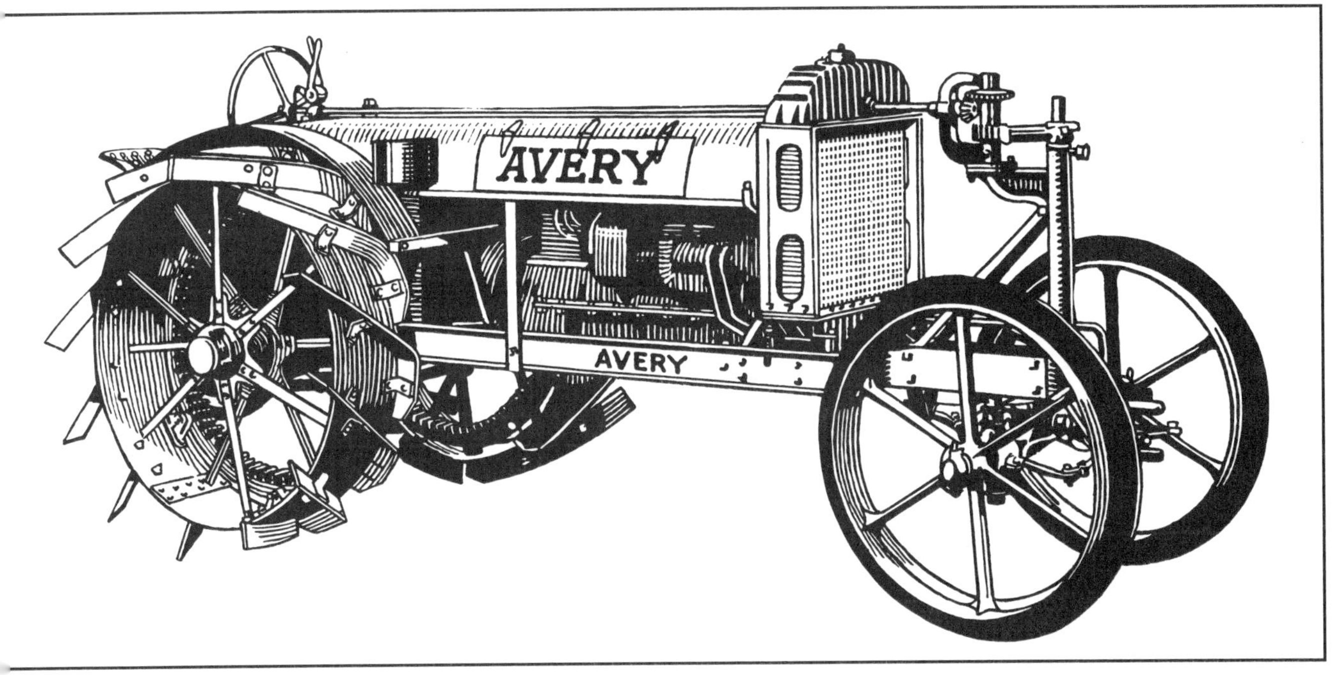

About 1919 Avery announced its Model C six-cylinder tractor. This small tractor likely replaced the earlier 5-10, which was a four-cylinder model. Although simple and compact, this model had sales totals less than expected, despite its attractive appearance.

The Avery Company supplied advertising cuts to its dealers, including this one that a dealer might put in the local newspaper. It illustrates mechanization of the hay crop, at least to some extent. Although two men are loading loose hay, the horses have been replaced with an Avery Model C tractor.

For orchards, groves, and vineyards Avery provided this Model C tractor with special orchard fenders and hood. This equipment protected tender branches from damage by the tractor, and likewise offered a degree of protection to the operator. Avery built this tractor for a short time, beginning about 1919.

The Avery Model C, when suitably modified, became the Avery Six-Cylinder motor cultivator. Numerous companies of 1920 designed and built motor cultivators in an attempt to mechanize this farm task. While this design has certain similarities to what would later become a row-crop tractor, the design simply didn't go far enough to achieve this ultimate goal.

Another modification of the Avery Six-Cylinder cultivator was this Avery Road-Razer one-man grader. It was essentially the same unit as the motor cultivator, except that the cultivating equipment was replaced with a blade mounted on a sub-frame beneath the tractor.

A unique tractor design was this Avery 15-30 model of 1923. Once again, the company attempted to bolster its eroding market with an entirely new model. Even though the 15-30 was comparable in design and appearance to many of its contemporaries, sales were disappointing. What had been a company with a net worth of $4 million a decade before, went into bankruptcy in 1924.

The Avery Track-Runner of the early 1920s was an ambitious, albeit short lived, attempt by Avery to get into the farm crawler business. Unfortunately, due to the under-design of various track parts, frequent repairs were necessary. This may have contributed to the very short production period for a most interesting crawler model.

This 1917 offering from Avery was likely the first tractor designed specifically for nursery work, and it probably holds the record as the first high-clearance farm tractor. A large funnel at the front of the tractor fed nursery stock into the tunnel beneath, protecting the growing stock from injury while cultivating, spraying, or other necessary tasks. Eventually, many different companies would offer high-clearance tractors for special applications.

Bates Machine & Tractor Company, Joliet, Illinois

The famous Bates Steel Mule emerged in 1918, remaining on the market for several years. This interesting design weighed less than 5,000 pounds, despite its rating of fifteen drawbar horsepower. Bates Steel Mule was a tradename well in tune with the times, especially if the term would be helpful in convincing a recalcitrant farmer that the "steel" mule would be much better than a "real" mule.

The track mechanism of the Bates Steel Mule is shown in this exploded illustration directly from a Bates catalog. Since there was constant wear on the pins and bushings, it was imperative that there be a means whereby these parts could be easily replaced when necessary. At the time, there were very few large machine shops, especially in rural areas. Repair work almost always was done on-site, and that meant out on the farm, a long way from the nearest garage. Note that even the rear drive sprockets were designed with a removable rim to lower replacement costs.

Bean Spray Pump Company, San Jose, California

Between 1918 and 1920, Bean manufactured this small tractor with a single front-mounted track drive. Its small size and low profile made it ideal for orchard and vineyard work. While this unique and very unusual design certainly had its advantages, the short production run of only two years would tend to indicate that sales were somewhat less than anticipated or desired. The Bean Track Pull was rated at six drawbar and ten belt horsepower.

Beltrail Tractor Company, St. Paul, Minnesota

Particularly between 1910-20, tractor designs ran the entire gamut. Every conceivable configuration was used; some were successful, and others failed completely. One of the unique designs of the period was the Beltrail tractor. Built in the 1917-20 period, this small tractor with 20 belt horsepower used a single crawler track. A 1917 advertisement notes that it had already undergone "nearly three years of severe testing."

Benjamin Tractor Company, Dallas, Texas

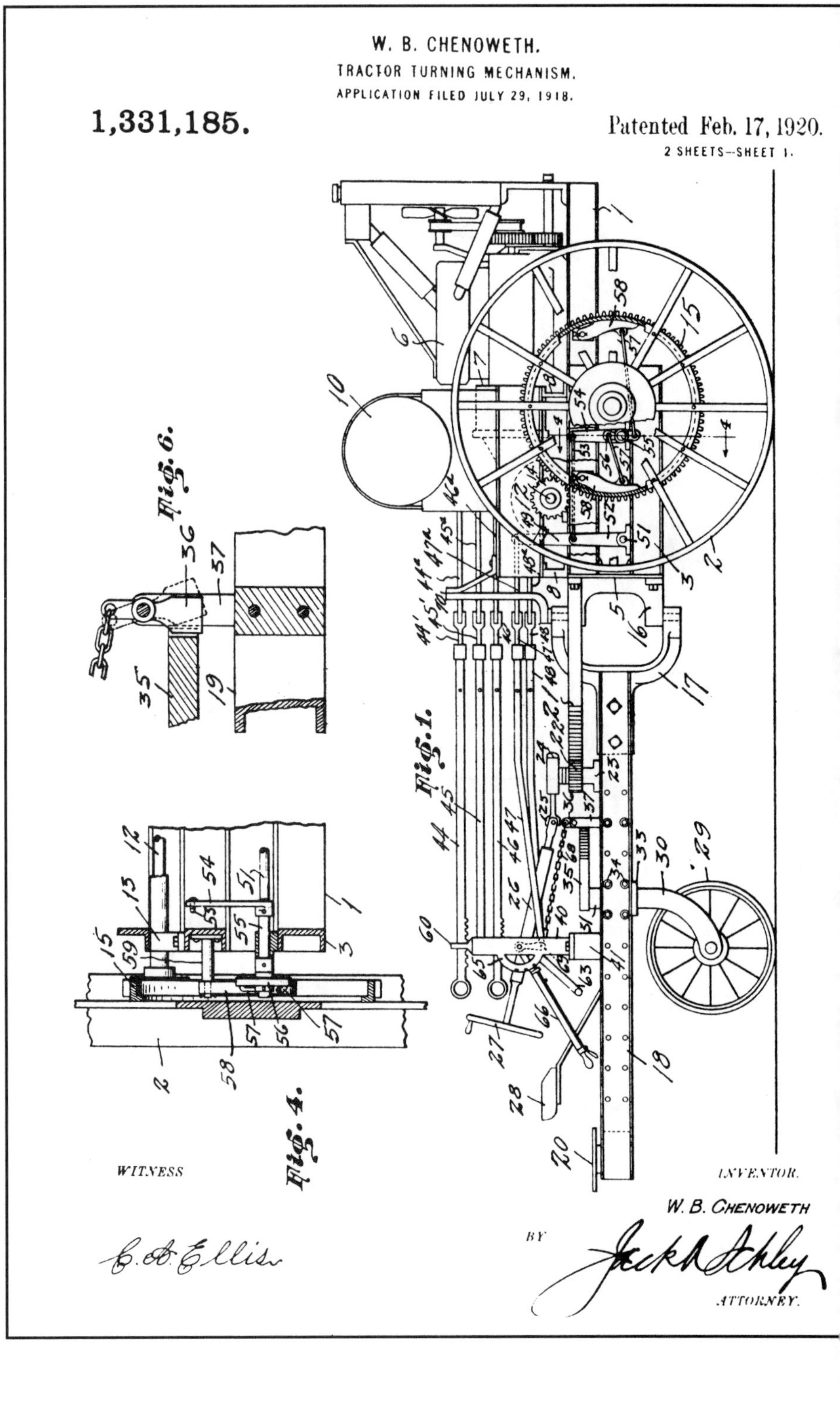

In 1918, W. B. Chenoweth filed this patent application for his unique tractor design. When it was granted in 1920, the Benjamin Tractor Company was organized. It managed to build at least a few tractors, and advertised the new machine in 1920 as the Big Ben. According to its advertising, "It will push, pull or operate any farm implement or belt driven machine used on a farm." The Big Ben was rated by the manufacturer as having twenty drawbar and forty belt horsepower; they also rated it as an "8-mule" tractor. After 1920, nothing more can be found concerning the Big Ben.

Boenker Motor Plow Company,
St. Charles, Missouri

Farm tractor design moved forward with amazing speed in the 1910-20 period. Literally thousands of individuals conceptualized the ideal farm tractor, and hundreds of them brought their designs to fruition. The Boenker Motor Plow of 1911 was one such design; it attracted considerable attention at the time, although it never attained wide popularity.

Boring Tractor Company, Rockford, Illinois

In 1917 Boring came out with this interesting and unique design. The two drive wheels were forward, with the engine set above. This was not unusual for tractors of the period. The unique feature was the small steering wheel at the rear of the tractor. The operator was seated directly over this steering wheel, and no doubt was grateful for the spring-mounted seat. Production of this tractor continued into 1922.

Bull Tractor Company, Minneapolis, Minnesota

When the Bull tractor came onto the market in 1913 it was an immediate sensation. Here was the first successful small tractor to be offered, and farmers bought them by the thousands. Old-timers in the tractor business found that farmers were far more interested in the small Bull tractors than in their heavyweight behemoths, and within the year, virtually every manufacturer, and those aspiring to be so, were attempting to build a "small" tractor. Bull led the industry in sales during 1914, but by 1919 the industry passed them by, forcing a merger with Madison Motors Corporation of Anderson, Indiana.

J.I. Case Company, Racine, Wisconsin

This company was founded by Jerome Increase Case in 1842 to build grain threshers. They were tremendously successful, and by the 1860s, Case threshers were widely known and used. On these foundations the company built its first portable steam engines in 1869, and from that point on, the company made rapid strides in the development of its own steam traction engine. Subsequently, Case went on to build far more steam tractions than any other firm.

With their tremendous success in the steam engine and thresher business, Case built their first tractor in 1892. It was not successful. At this point, reliable ignition systems were virtually non-existent, and atomizing liquid fuels with carburetors and mixers was poorly perfected. Despite extensive experiments, the single hand-built model was finally pushed into the corner of a warehouse, and was eventually scrapped.

Case reentered the tractor business in 1911 with a huge tractor having thirty drawbar and sixty belt horsepower. It was intended to compete with other big tractors of the day, but the 30-60 never gained the popularity of some of its competitors.

The following year, 1913, Case introduced its 20-40 tractor. Unlike the 30-60, this new model was immensely popular, perhaps because it was easily adaptable to average farms, and, of course, was far less expensive. Case gradually expanded their tractor line, but their most unique tractor appeared in 1915. This was the 10-20 model; it was a three-wheel design quite in keeping with the rage of the day. However, the following year Case began offering a new series of tractors using a crossmounted engine. Eventually the Case crossmount line became quite extensive, ranging all the way from ten to forty drawbar horsepower.

The company was named "J.I. Case Threshing Machine Co." in 1880. When Massey-Harris bought out the J.I. Case Plow Works in 1928 (there was never any corporate connection between the two Case companies), it sold the "Case" name to J.I. Case T.M.C. At this time "Threshing Machine" was dropped from the corporate name—from then on it was the J.I. Case Company.

In the late 1920s, Case abandoned the faithful crossmount designs and came out with an entirely new series of tractors of unit frame design. The Model C and Model L Case tractors were the backbone of the new line; in the opinion of many, these were among the finest tractors ever built by J.I. Case Company.

During the 1930s and 1940s Case continued with its proven designs, but by the 1950s it was time for a new line of tractors. Very stylish and modern for their time, these new models even included the Model 800 with the Case-O-Matic drive, a torque converter system that was immensely popular.

Late in 1984 came the Case-IH merger in which Case acquired International Harvester Company. From this combination of companies, the Case-IH line remains as a dominant force in the tractor industry.

J.I. Case came out with its first tractor in 1892, but it never got past the first experimental model. Nearly two decades went by until the company once again entered the tractor business with this huge 30-60 model in 1911. It was gigantic in every respect, but was much like many other tractors of the time. A modified and improved version of the 30-60 followed in 1912, with this tractor remaining on the market until 1916.

The popular Case 20-40 tractor first appeared in 1912. Production apparently ended in 1919, although advertisements for the 20-40 appeared as late as 1920. Although the 20-40 was modified slightly during its production run, it retained most of its salient features throughout the years. This view illustrates and describes many features of the 20-40, and also shows the two-speed transmission system with its massive open gears.

In 1915 Case launched its little 10-20 tractor, the company's first venture into the small tractor arena. By this time, Bull Tractor Company at Minneapolis was dominating the market with its three-wheeler, and many other companies, including Case, followed suit. Fortunately for farmers, this fad didn't last but a few years. Unfortunately for the manufacturers of the three-wheel design, huge sums of money were spent on a tractor that farmers were ready to abandon as soon as something better arrived on the scene.

By 1916 J.I. Case had developed a small tractor using a unit frame design and a crossmounted engine. The great selling point of the crossmount was that with this arrangement, no bevel gears were needed; all power to the drivewheels was transmitted through spur gears. Additionally, the Case crossmounts proved to be extremely rugged and reliable. This, together with the sterling reputation of the Case organization, made the Case crossmounted tractors tremendously popular. Many different styles and sizes were built until the company changed over to the unit frame design in 1929-30.

In the 1929-30 period Case introduced its Model L and Model C tractors. They were virtually identical except for size and power level. The Model L was usually rated at forty belt horsepower, with the Model C about half that size. The Case-built four-cylinder engine proved itself from the very beginning, and a great many Model L tractors found their way to the farm. This one is belted to a threshing machine, probably on a hot July afternoon. While those "good old days" of threshing retain a certain mystique for the

This typical threshing scene shows a Case Model L in the distance, belted to a Case threshing machine. A long drive belt was needed. With a short belt, it was very difficult for the team and bundle wagon to swing in past the tractor and get close to the feeder on the thresher. In addition, a long belt had more weight and a better grip on the tractor drive pulley and the cylinder pulley. A crossed or twisted belt was always desirable. A few tractors necessitated the use of a straight belt, and this was particularly troublesome with a cross wind; the wind would drift the belt off the pulleys. Case continued building the Model L, and later the Model LA, into the 1940s. Eventually though, other Case models took their place.

The Model CC Case was the first row-crop tractor from this company. Introduced in 1929, this model launched an entire series of row-crop tractors, and the basic design remained in production until the early 1950s. Numerous attachments were available, including cultivators for virtually any purpose. Case went on to build many innovative tractors in the 1950s, followed by even more models up to the present time. As with many of the tractors illustrated in this book, additional references are noted in the bibliography at the end of this volume.

J.I. Case Plow Works, Racine, Wisconsin

In 1913 J.I. Case Plow Works came out with the Wallis Cub. This was the first American tractor to use the unit-frame design. The unit frame in the Cub was rolled from boiler plate, and eliminated the need for a separate steel frame to support the engine and drivetrain.

During 1915 the J.I. Case Plow Works introduced its Wallis Cub Junior tractor. This model epitomized the unit frame design, and in addition, all gears were enclosed, with live axle shafts to each rear wheel. Despite claims by Henry Ford that his Fordson tractor was the first with a unit frame, the Cub Junior predated the Fordson by about two years. However, it must be conceded that the Fordson was the first tractor with the entire transmission case being a single casting.

For a short time, apparently for the 1919 crop season, Case Plow Works offered its Case Motor Cultivator. This tractor closely approached the row-crop design adopted by the Farmall from International Harvester only a few years later. Had the Case Plow Works engineers envisioned the potential for this design, the company's future might have been much different. As it turned out, the company was bought out by Massey-Harris in 1928.

After the Wallis Cub came the Model K tractor in 1919. This design closely followed the unit frame concept in the earlier Cub tractor, along with numerous modern features such as completely enclosed gearing in a constant oil bath. Roller bearings were used extensively to minimize friction—this at a time when many tractors still used plain babbitt bearings. In many respects, the Model K was far ahead of its time.

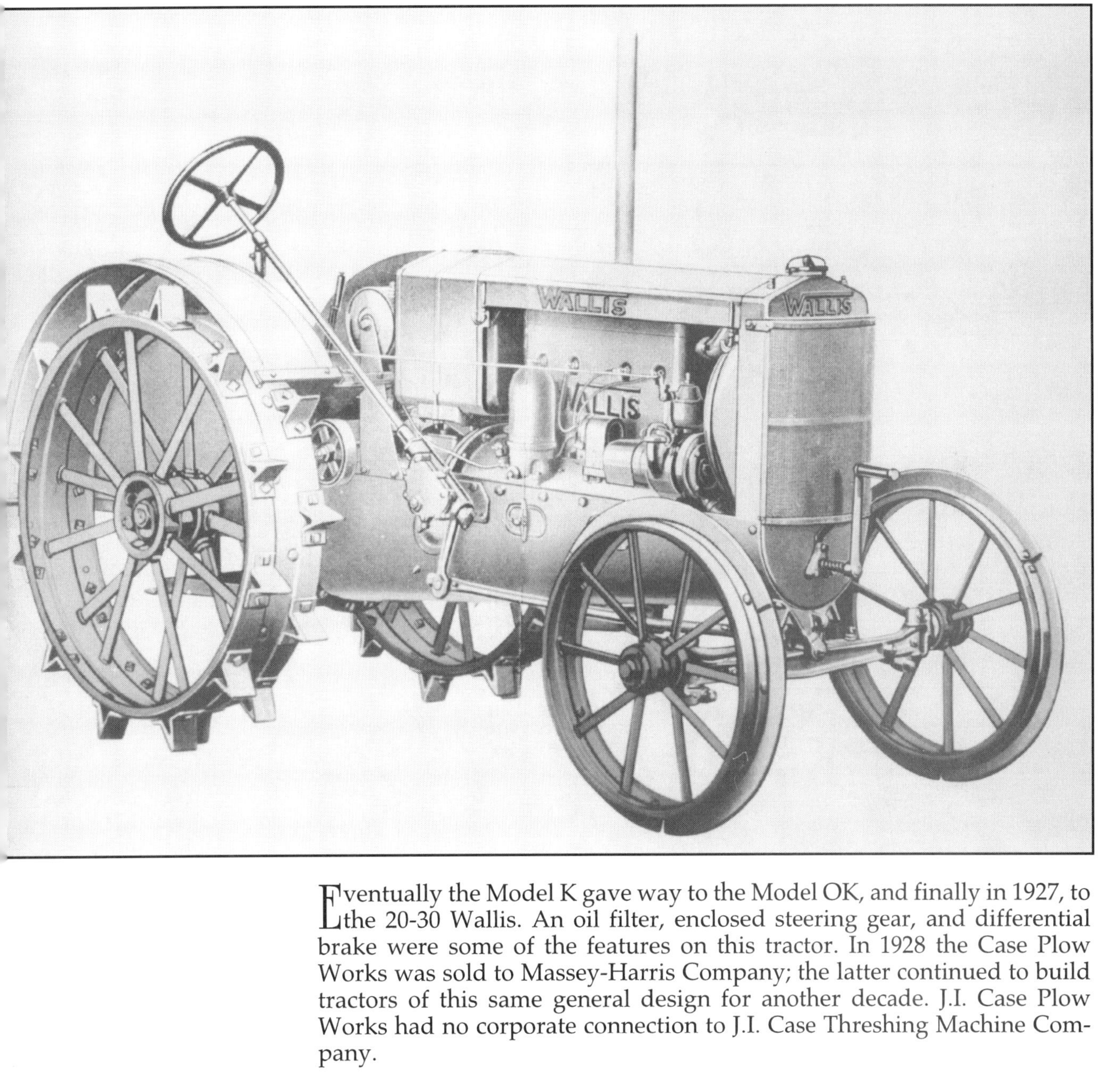

Eventually the Model K gave way to the Model OK, and finally in 1927, to the 20-30 Wallis. An oil filter, enclosed steering gear, and differential brake were some of the features on this tractor. In 1928 the Case Plow Works was sold to Massey-Harris Company; the latter continued to build tractors of this same general design for another decade. J.I. Case Plow Works had no corporate connection to J.I. Case Threshing Machine Company.

Caterpillar Tractor Company, Peoria, Illinois

Crawler tractors developed in parallel with wheel-type models. Of all manufacturers of crawler tractors, Caterpillar rose to the top almost from the beginning, and has remained in a dominant position ever since. In the contest between quality and price, Caterpillar opted for quality, building its machines to the most exacting standards. Caterpillar also made history with its first diesel-powered crawler in 1929, placing it on the market late the following year. Other manufacturers have come and gone, but the Caterpillar name remains dominant in the tractor industry.

C.L. Best Co. introduced a special orchard crawler in 1915. Known as the "Humpback," this model was modified so as to lower the height profile that was imperative for work in orchards. The track drive is obvious in this photograph and shows a small drive pinion mounted above the rear idler wheel. At this point, a front tiller wheel was still in use. Eventually a similar track design would appear, but this time on modern Caterpillar tractors.

In the 1920s the Caterpillar Sixty tractor appeared; it remained in production until the early 1930s. This big crawler tractor was an instant hit with contractors and road builders. A substantial number were also used in large farming operations. The Sixty used a big four-cylinder engine, and was capable of pulling well over six tons at the drawbar. A substantial number of Caterpillar Sixty tractors still exist, with most of them receiving tender care from enthusiastic tractor collectors.

In 1925 Holt Mfg. Co. and C.L. Best merged to form the Caterpillar Tractor Co. Prior to that time, Best developed a rear-mounted winch for its large crawlers, with this one mounted on a Best Sixty tractor. For many applications this was an immensely useful device; Caterpillar later developed it even further. From this basic design came the double-drum cable control with which Caterpillar controlled its huge wheel scrapers.

The Caterpillar Diesel Fifty tractor emerged in the early 1930s and, like other Caterpillar Diesels, became quite popular. The overall design of the Diesel Fifty remained relatively stable for many years; this is indeed a tribute to the quality of the original design. By the mid-1930s the Caterpillar Diesel engine was outselling the company's gasoline powered designs,

By 1937 Caterpillar was offering diesel tractors in five different models. This was the beginning of the D-Series tractors, such as this D-4 model. For Caterpillar, the D-Series tractors marked the beginning of an era of dominance in the crawler tractor industry that continues to the present time. Over the years the company has also developed a wide range of construction equipment, as well as providing OEM engines for a wide range of applications.

Cleveland Tractor Company, Cleveland, Ohio

Hay mowing, raking and loading are speeding up with the Cletrac. No time out to "blow" it as with teams

Hillsides are always cultivated on the Cletrac-operated farm

In 1917 Cleveland Tractor Company introduced its Cleveland Motor Plow to the market. Within a short time the company was offering various implements for use with its new Cletrac models, and pioneered the use of a front-mounted cultivator. The little 12-20 model shown here remained on the market into the early 1930s. At the time, Cleveland was a major player in the crawler tractor industry.

A salient feature of the Cletrac design was its specially designed planetary gear system. Using this design, power was delivered to both tracks, even when turning. Conversely, other designs disengaged one track when turning with consequent slippage. Another unique feature of the early designs was the use of an automotive-type steering wheel rather than individual steering levers.

By the 1930s Cletrac was offering an extensive line of crawler tractors. Although a great many found their way into construction and road work, the Cletrac was ostensibly sold as a farm tractor, particularly in the smaller sizes. To this end, a rear-mounted belt pulley was regularly furnished. As otherwise noted, Cletrac heavily promoted their smaller tractors for cultivating, and went so far as to develop a "quick-on, quick-off" attachment system. In-field servicing was fairly simple, since the track rollers could be disassembled and repaired by the average shop.

In the mid-1930s Cletrac offered its first diesel-powered tractors, although these were available only in the larger models. The company remained very aggressive in the crawler tractor market, despite the blight caused by the Great Depression of the 1930s. Oliver Corporation took over the Cletrac line in 1944, and continued building crawler tractors until all production ceased in 1965.

Coleman Tractor Company, Kansas City, Missouri

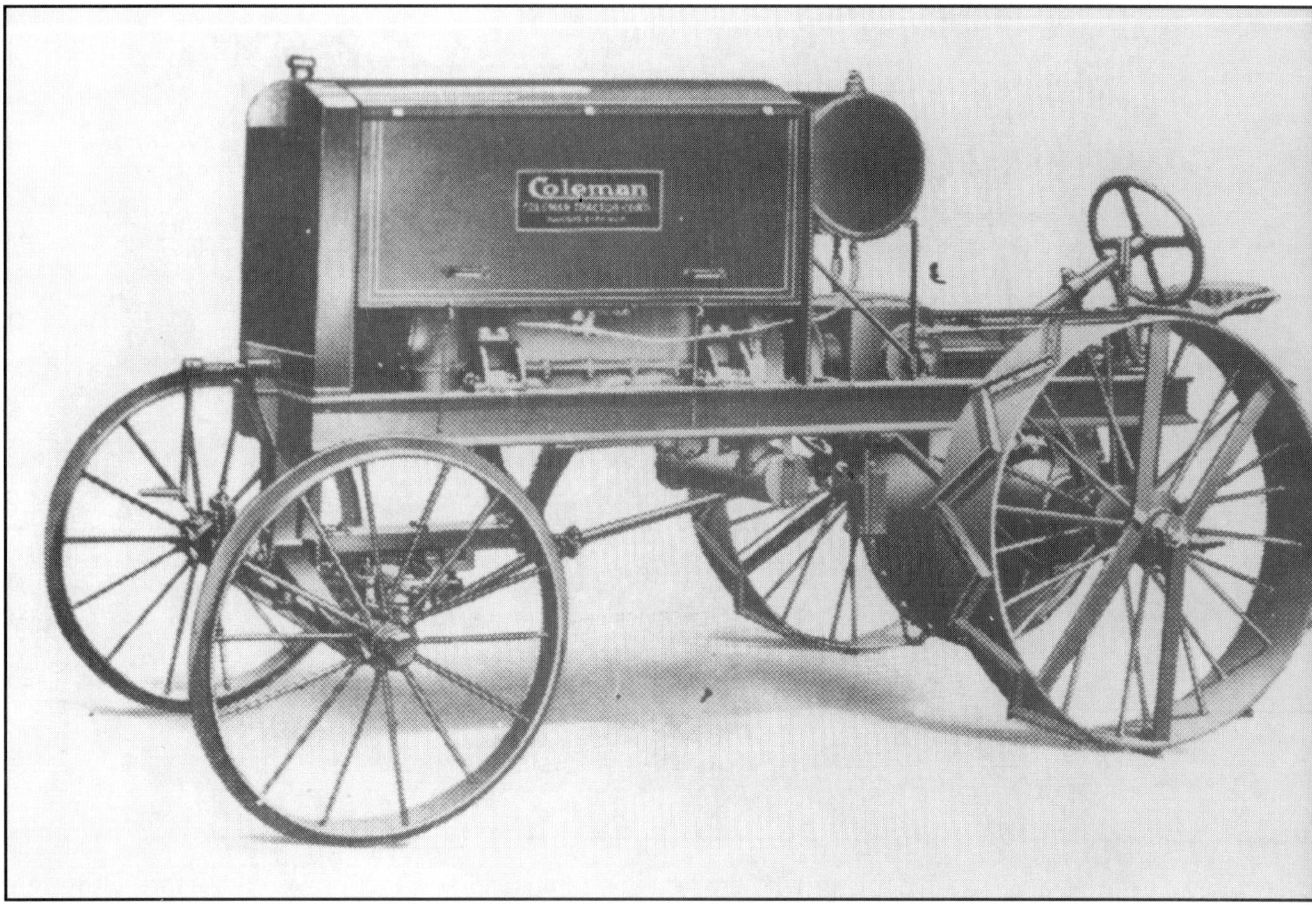

Beginning in 1918 and continuing into 1920, Coleman built this interesting tractor model. Although the engine was completely hooded, and well protected, the operator had only the comfort of a steel seat and no fenders. Thus, when doing field work in a cross wind, the operator was subjected to all the dirt, grit, and other debris from the drivewheels. A study of the photo will reveal some interesting design features, including the reinforcing bars from the live axle to the rear wheel rims.

Columbus Tractor Co., Columbus, Ohio

In 1918 the McIntyre Mfg. Co. was organized at Columbus to build this "Farmer Boy" tractor. Later that year the company was reorganized as Columbus Tractor Co. While it was only built for a short time, the "Farmer Boy" displayed many design features that would soon be incorporated into row-crop tractor designs. However, this one used a single drive wheel, and its huge open gears were rapidly approaching obsolescence.

Comet Automobile Company, Decatur, Illinois

Comet built automobiles from 1917 to 1922. Also, during 1919 and 1920, the firm got into the tractor business with this unique design. Like several other tractors of the period, the Comet was built up of various OEM parts. The engine was by Erd Motor Company, and the transmission came from Foote Bros. Company. Despite its attractive appearance, the Comet, like many others, made little headway in an extremely volatile and competitive market.

Common Sense Gas Tractor Company, Minneapolis, Minnesota

The interesting Common Sense tractor first appeared in 1915, and remained on the market until 1919. Its most distinctive feature was the use of a V-8 engine; at the time, few cars were equipped with a V-8, so the idea of putting a V-8 in a tractor was truly sensational. The company advertised that it was a "1-Man, Boy or Girl Outfit." This alone was a good sales point in an era when most tractors required an expert engineer, and sometimes an assistant.

Continental Cultor Company, Springfield, Ohio

About 1927 the Continental Cultor appeared as an answer for those wanting a small all-purpose tractor. Intended for small acreages, the Continental used a Ford Model T engine, and this was a great advantage, since parts were readily available. However, the coming of small row-crop tractors effectively brought the days of the universal tractor to an end. Farmers were more than ready to buy a tractor that was capable of all their farming needs, rather than have separate units for each job.

Continental Tractor Company, Continental, Ohio

This interesting tractor-plow of about 1917 placed the operator directly over the plows in much the same way as a horse-drawn gang plow. Leg support is totally absent, except, perhaps, for the plow beams. All controls were within easy reach of the operator. The unusual location of the belt pulley made it necessary for the drive belt to pass directly alongside the tractor operator. Except for a couple of advertisements, very little is known of this company or its tractors.

Craig Tractor Company, Cleveland, Ohio

The unique feature of the Craig was its unusual rear wheel design. Outside of this feature, the Craig was much like other tractors of the 1919-21 period when the Craig was on the market. Literally hundreds of tractor manufacturers sprang into bloom in the 1915-20 period; the vast majority never saw anything approaching success in their ventures.

Curtis Form-A-Tractor Company, Chicago, Illinois

By 1918 there were probably fifty or more companies offering tractor attachments for automobiles. One of these was the Curtis Form-A-Tractor, priced at $350. As illustrated here, the rear wheels remained in place, with leather straps buckled around the rims to transmit power to the traction wheels. The company claimed the changeover from car to tractor and back again could be done in five minutes. Supposedly, the idea was to run the Form-A-Tractor in the field during the week, drop off the attachment for Saturday night and church on Sunday, then back to the field again.

Dart Truck & Tractor Corporation, Waterloo, Iowa

Dart was a company owned in part by the famous Bill Galloway of gas engine fame. Although this model was rather attractive for a 1920 tractor, it appears that when Galloway was forced into bankruptcy in 1920, the tractor also fell into the abyss. While little is known of the Dart, this one featured advanced designs such as a spring-mounted front axle.

Dauch Mfg. Company, Sandusky, Ohio

Dauch's Sandusky tractor is one of those elusive models for which very little information is available. The company began building tractors about 1912, continuing until 1920. This one, introduced in January of 1917 demonstrates streamlined hood and radiator lines quite comparable with contemporary automotive design. The Sandusky was built almost entirely in-house, with Dauch even manufacturing its own engines for its tractor models.

Dayton-Dick Company, Quincy, Illinois

Dayton-Dick Company was organized in 1915, and two years later, this 12-18 model appeared; it was built as late as 1921. The Leader tractors from Dayton-Dick remained in the marketplace until 1924, and this despite a postwar depression and the most brutal competition within the farm tractor industry.

Deere & Company, Moline, Illinois

In 1892 John Froehlich built his first tractor in the small village of Froehlich, near Monona, in northeastern Iowa. Its initial success led to the founding of the Waterloo Gasoline Traction Engine Co. the following year at Waterloo, Iowa. While this early design was reasonably successful, commercial production was never begun. Instead, the company concentrated its efforts on stationary engines for some years, finally resuming serious tractor experiments about 1910.

By 1912 Waterloo Gasoline Engine Company was back in the tractor business, billing the heavy four-cylinder model shown here as the "Waterloo Boy One Man All-Purpose Tractor." Little is known about this early design, although it obviously was equipped with a four-cylinder engine. Production of this model apparently lasted for only a year or two.

In the 1912-13 period, the Waterloo Boy tractor took on an entirely new appearance, using a two-cylinder opposed engine and rear-mounted plows. As with the 1912 model, very little information has been found on the tractor shown here, but obviously much of the design work was carried over into the later Model R and Model N Waterloo Boy tractors.

Waterloo Boy Small Farm Tractor

The original 12-24 Waterloo Boy tractor first appeared in 1914, and was built until 1916; during the latter year it became known as the Model R tractor. It remained in production until 1919. Deere & Company bought out the Waterloo Gasoline Engine Company in 1918. In so doing it instantly gained a proven tractor design, with roots in the tractor business going back to the Froehlich tractor of 1892.

Waterloo Boy Model N tractors were built from 1917 until 1924. This model differed slightly from the Model R, in that it used a two-speed transmission, while Hyatt roller bearings replaced many of the plain babbitt bearings found on the Model R. Rated at twelve drawbar and twenty-five belt horsepower, this was the first tractor to be tested when the University of Nebraska's tractor tests began in 1920.

One of the interesting aspects in researching the history of the farm tractor is the advertising used to induce farmers to purchase this or that machine. This illustration shows a farmer pulling a tandem disc with a roller or clod buster behind on land as level as a table. Seldom were conditions so pristine as this, yet the message came through that only one person was needed to accomplish all this work with no stops needed to "wind" or water the horses. The advertisement is from the February 1920 issue of *Power Farming Magazine*.

Originally known as the John Deere 15-27, this tractor soon was dubbed the Model D. Regular production began in 1924, finally ending in 1953. Although the Model D saw numerous changes during its production run, the basic design remained the same, with the Model D becoming the backbone of the extensive John Deere tractor line. With the Model D, John Deere developed the two-cylinder tractor to its ultimate, and all the while creating one of the most popular farm tractors ever built.

In the 1928-35 period, Deere built a "little brother" to the Model D. This small tractor, rated at ten drawbar and twenty belt horsepower, was also known as the "GP" which was an acronym for "General Purpose." The GP used an arched front axle to accomodate its new line of three-row implements; this was the company's first approach to the row-crop tractor problem. Instead of the valve-in-head engine used in the Model D, the GP tractor used an L-head engine.

Deere produced its first row-crop design with the GP Wide-Tread model. Built in the 1929-33 period, it was the immediate forerunner of the famous Model A, built from 1934 to 1952. The Model A was the first tractor with an adjustable rear wheel tread, and was also the first to use a one-piece transmission case. Individual differential brakes and a power lift system were other major features.

A styled hood and grille were the most noticeable changes on the new John Deere line introduced in 1939. Over the long production run of the two-cylinder line, Deere introduced many different sizes and styles of tractors, even building its first two-cylinder diesel, the Model R, in 1949. Eventually though, the demand for increased power levels brought about the need for four- and six-cylinder engines to achieve that goal.

The first BO-Lindemann crawlers were produced in 1939. The Lindemann folks fitted crawler tracks to a Model BO tractor, thus creating a small two-cylinder crawler. In 1949 the John Deere MC crawler shown here made its first appearance, with production continuing into 1952. From these early beginnings came today's extensive line of John Deere crawler tractors.

Denning Tractor Company, Cedar Rapids, Iowa

Joseph Denning was a prolific inventor, and developed this small machine about 1913. Built to be small and lightweight, it was powered by a small two-cycle, 12-hp engine. Despite its simplicity, this model never reached the hoped-for success of its inventor. Subsequent models were of conventional four-wheel design, but the company went bankrupt in 1919 and was sold to General Ordnance Company; the latter went into receivership in 1922.

Denning's earliest tractor designs were built over a frame of steel pipe. The entire tractor was intended to be as lightweight as possible, and this in itself was a welcome change for tractors of 1913. Another innovative feature was the use of two-forward speeds, a concept that would soon gain favor, but which was not evident on the great majority of tractors built at the time.

Depue Bros. Mfg. Company, Clinton, Iowa

Built during the 1918-24 period, the Depue looked somewhat like a truck, but was designed as a four-wheel-drive tractor. As is obvious in this advertising photograph, the Depue was completely hooded, even including the transmission case. The operator's cab apparently came as standard equipment, and further evidence of automotive thought is demonstrated by the placement of the throttle and spark control levers on the steering column. Few specific details can now be found on this interesting design.

Detroit Engine Works, Detroit, Michigan

In the early 1900s Detroit began building two-cycle marine and stationary engines. These gained a fine reputation, and in 1914, or perhaps earlier, the company made a small attempt to enter the tractor business with its Wadsworth tractor. It was, of course, powered by a Detroit twelve horsepower engine, and was made light and simple to lower the price and take away some of the mystery for apprehensive farmers. However, the Wadsworth made little or no impact, and the entire effort was abandoned shortly after it began.

Detroit Tractor Company, Detroit, Michigan

This company appears at Detroit, Michigan, as well as at LaFayette, Indiana. Introduced in 1914, the company claimed this tractor was the result of fifteen years of experience in the tractor business. It was controlled by lines or reins, and probably was intended to convince those reluctant to part with horses that this machine needed no rest or water stops; in other words, it was the same as driving a team, except that an engine replaced muscle and brawn. It was built for a short time and then disappeared from the market.

Dill Tractor Mfg. Company, Little Rock, Arkansas

Designed for a specific need, the Dill tractor used extremely wide wheels to gain maximum flotation. This was needed in swampy rice-growing areas, and the Dill tractor was a 1920s attempt to mechanize this farming specialty. While it appears that the Dill was locally popular, there are no indications that it was marketed to any extent outside of rice growing regions.

The Dill tractor used a very long wheelbase, and as can be seen in this photograph, the tractor was designed so that certain machines could be mounted directly to the tractor chassis. In this instance, a grain binder is mounted directly to the tractor; it was driven through a power shaft from the tractor engine. Note also the extremely wide wheel tread; this provided the necessary flotation for working in wet and swampy ground.

C.H.A. Dissinger & Bro. Company, Lancaster, Pennsylvania

Already in 1905, Dissinger was offering its Capital gas tractor to the market, the Dissinger brothers having gained considerable background with Otto Gas Engine Works at Philadelphia in prior years. The company appeared in the various tractor directories as a tractor manufacturer up to the 1920s, but eventually disappeared from the scene. Although bits and pieces of this tractor have similarities to others of the day, it appears that the Capital was built entirely by Dissinger.

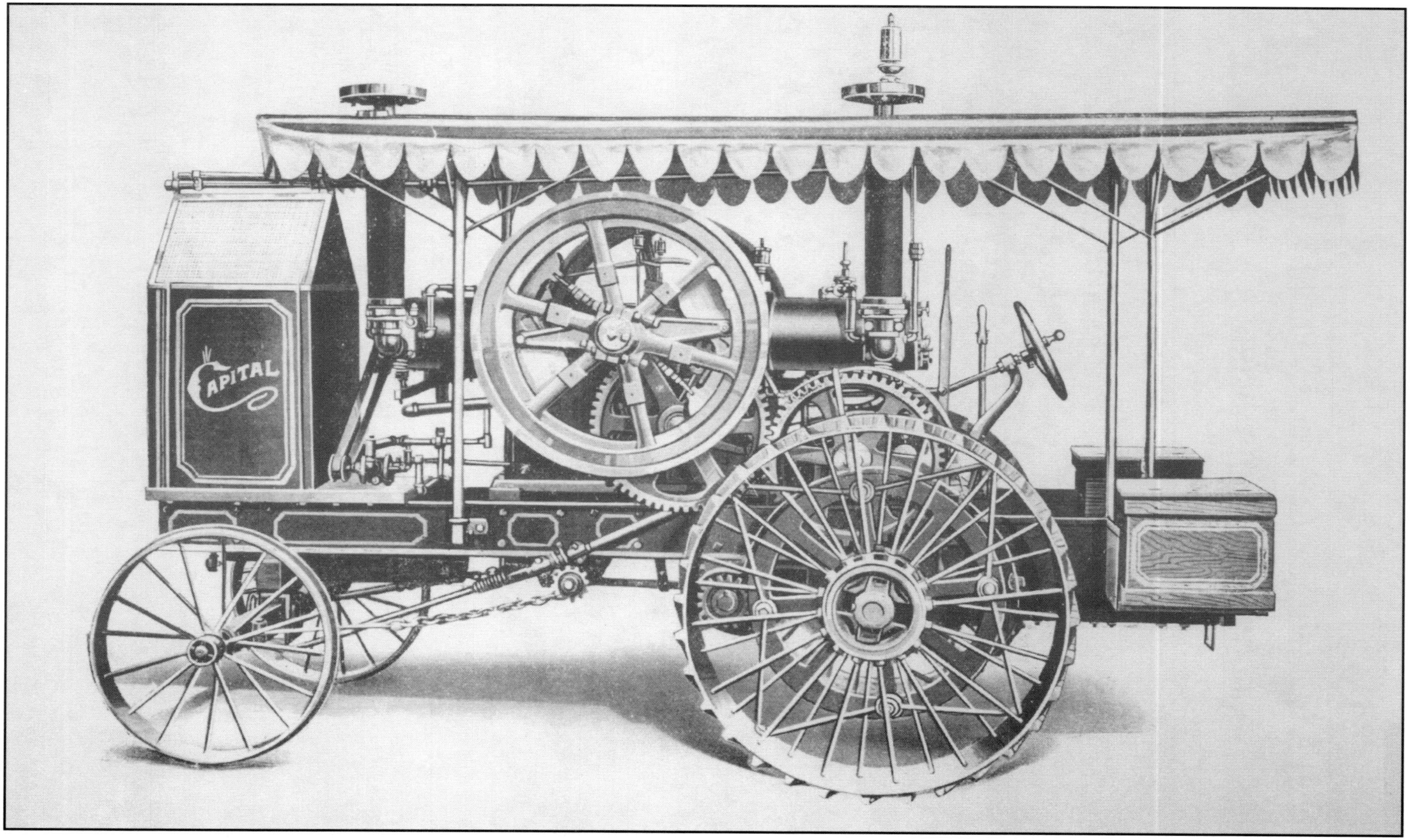

A Dissinger catalog of about 1910 indicates that the Capital tractor shown here was the forty-five horsepower model; its two-cylinder engine had a 10 1/2 x 15 inch bore and stroke. Similar in design, the seventy-five-horsepower tractor used a two-cylinder engine with a bore and stroke of 12 x 16 inches. The company also offered a four-cylinder, eighty-horsepower tractor; it carried bore and stroke dimensions of 10 x 15 inches. The tractor shown here sold for $4,400, while the seventy-five-horsepower model list-

H.C. Dodge, Inc., Boston, Massachusetts

The advertisements for the Sprywheel garden tractor appeared in 1921 and 1922. Outside of a few advertisements and announcements, it doesn't appear that the Sprywheel was marketed for any length of time. It was a lightweight model, and to that end, the company equipped the Sprywheel with a two-cycle engine. Although the majority of research into the history of the tractor has been confined to the farm tractor, these small garden tractors were also an important part of farm mechanization.

Eagle Mfg. Company, Appleton, Wisconsin

Eagle entered the tractor business as early as 1906, but little history can be found concerning its early years in the tractor business. However, the company emerged with the design shown here about 1917, with selected sizes remaining available into the late 1930s. The Eagle was heavily built, and was thoroughly reliable and dependable. In addition to the heavyweight designs, Eagle also began building various unit frame designs by 1930, and continued with these until production was suspended during World War II.

Elderfield Mechanics Company, Port Washington, Wisconsin

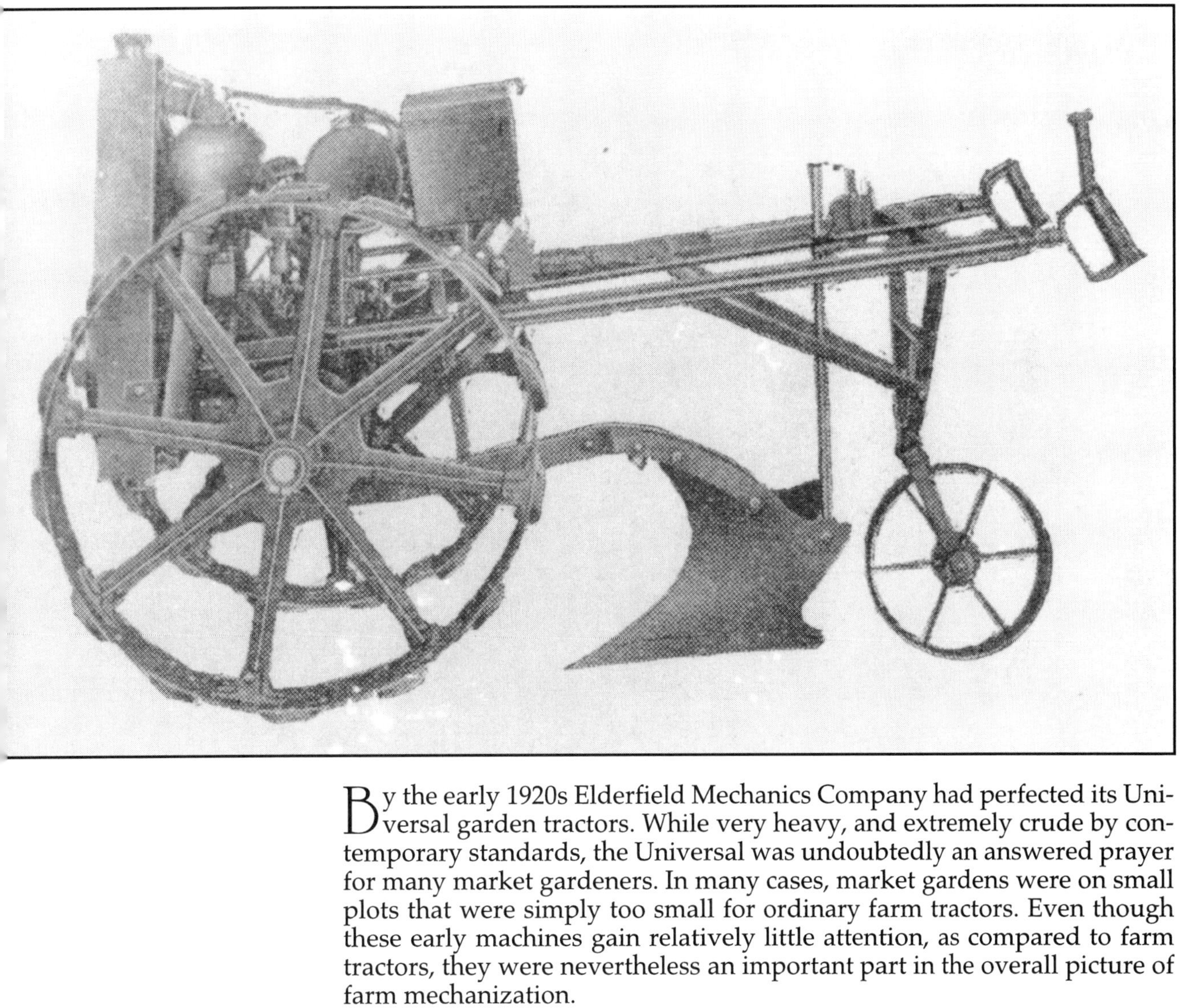

By the early 1920s Elderfield Mechanics Company had perfected its Universal garden tractors. While very heavy, and extremely crude by contemporary standards, the Universal was undoubtedly an answered prayer for many market gardeners. In many cases, market gardens were on small plots that were simply too small for ordinary farm tractors. Even though these early machines gain relatively little attention, as compared to farm tractors, they were nevertheless an important part in the overall picture of farm mechanization.

Electric Wheel Company, Quincy, Illinois

This company was formed in 1890 to manufacture all sorts of metal wheels. By 1908 Electric Wheel was offering a "traction truck," which was essentially a tractor chassis to which a farmer could mount his own gasoline engine for an instant "tractor." By the early 1920s Electric Wheel was offering this All-Work tractor. Its crossmounted engine eliminated the need for bevel gears, and, in fact, the crossmounted design was very popular for several decades for this very reason.

Elgin Tractor Corporation, Elgin, Illinois

The Elgin was a 1917 takeoff from the earlier Waite tractor built in Chicago, Illinois. Late in 1917 the company relocated to Piqua, Ohio, and continued in business there for a few years. The Elgin was a most interesting design for 1917, as evidenced by the completely enclosed hood, elevated air stack, and automotive steering. The full fenders were indeed a benefit for the operator. However, by 1920 farmers were demanding even smaller and more compact tractors. The ultimate goal was a tractor that could do any job on the farm.

Emerson-Brantingham Company, Rockford, Illinois

Emerson-Brantingham Company was a major player in the farm equipment industry by 1910, and two years later, it entered the tractor business primarily through the buyout of several tractor and implement manufacturers. This 1918 model was of lightweight design, and evidenced a full hood, half-fenders, and other attractive features. Attention was also paid to ease of service; in this illustration is shown the simplicity of removing a complete piston and rod assembly with the removal of a single hand hole plate. Despite the size of the company, fierce competition, a postwar depression, and other factors forced the company to sell out to J.I. Case Company in 1928.

In 1916 Emerson-Brantingham announced its Model L farm tractor. Rated at twelve drawbar and twenty belt horsepower, this model used a single rear drive wheel quite similar to other designs then in vogue. However, this design retained its popularity for only a few years. Farmers of 1917, the last year of production for the Model L, were getting their first look at the little Fordson and other small tractors. The result was that most of the three-wheel designs disappeared from the scene, along with many other tractors of considerable merit.

Fairbanks, Morse & Company, Chicago, Illinois

Few, if any, American manufacturers did more to perfect and popularize the gasoline engine than Fairbanks-Morse. In fact, this company aggressively pursued the engine business and built more engines than any other manufacturer. In 1911 Fairbanks-Morse turned its attention to tractors built on a chassis of its own design and using its own engine. The 15-25 model shown here was fairly popular for a few years, and a few still exist. The company also built a huge two-cylinder model rated at thirty drawbar horsepower.

Fig. HE1535—15-25 H. P. Tractor, left side

A 1913 advertisement illustrated the various features of the Fairbanks-Morse tractors; these are self-explanatory. Fairbanks-Morse was a huge company, and had the financial and shop resources for research and development of its products, including its tractors. The F-M Engineering Department had few equals in the engine and tractor industry, and the F-M shops were equipped to manufacture virtually anything desired, from formed steel shapes to castings of any size or description.

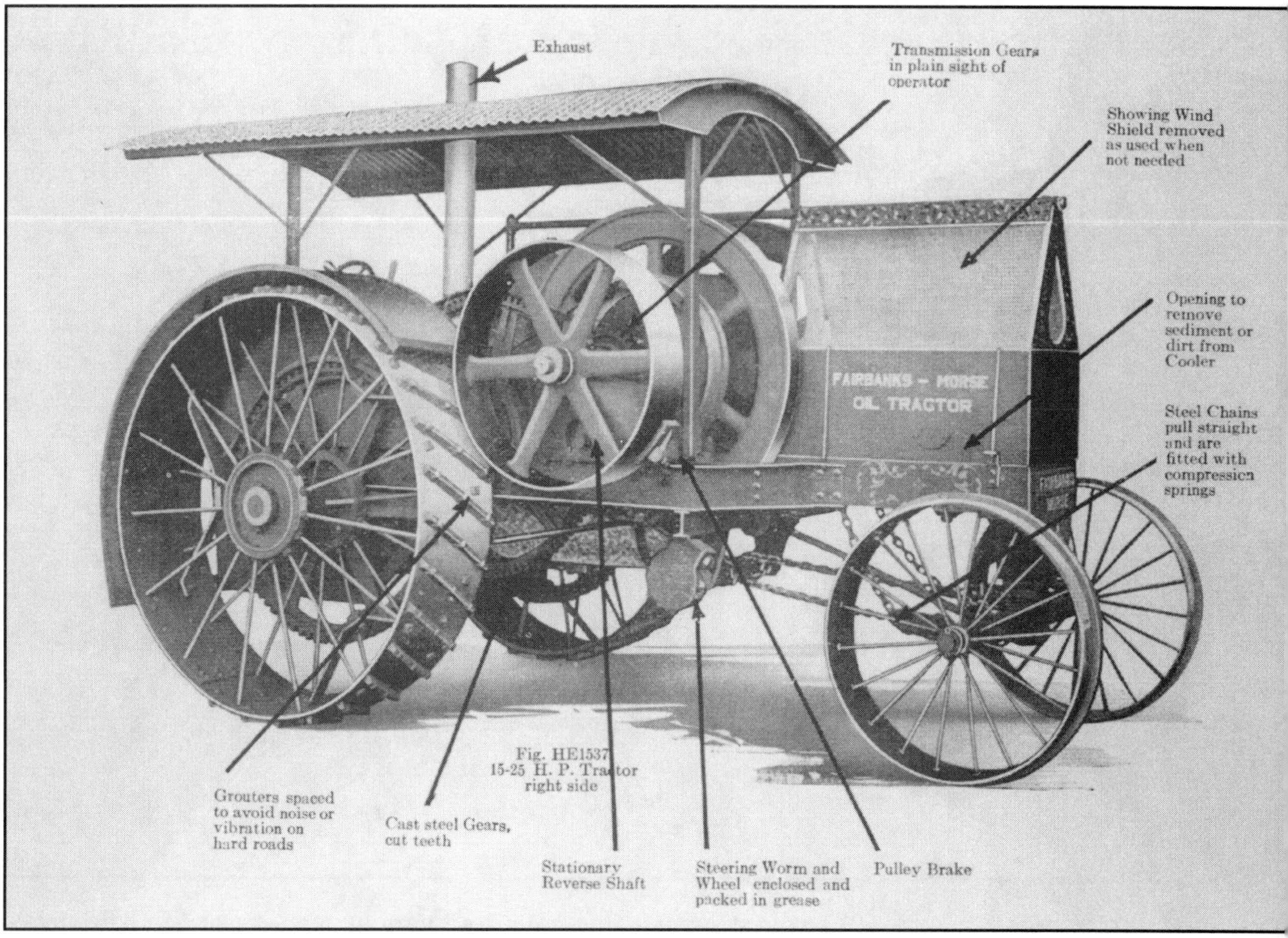

For reasons unknown, the Fairbanks-Morse tractors never attained great popularity. From the standpoint of quality and reliability, it had few equals. On the other hand, Fairbanks-Morse built more stationary gas engines than any other American manufacturer; even the gigantic International Harvester Company withered against the onslaught of F-M engines. Fairbanks-Morse had an excellent dealer organization as well. Whether it was a lack of marketing, or a lack of commitment on the part of the company is unknown. For whatever the reasons, the Fairbanks-Morse tractors were sold for only a few years, and then in very small numbers.

Farm Tractor Company, Fond du Lac, Wisconsin

About 1918 the Fond du Lac Tractor Corporation appeared as a manufacturer of "tractors, tractor attachments, and kerosene throttling governor engines." Virtually nothing is known of the company, except for this 1919 advertisement showing a conversion unit whereby a rather attractive touring car has been converted into a small tractor. The dearth of information of this company inevitably leads to the conclusion that this enterprise was of short duration.

Farmers Union Central Exchange, St. Paul, Minnesota

In the late 1930s, Co-op tractors appeared. The tractor shown here was equipped with a Chrysler engine, and could be serviced at any of some 12,000 Chrysler, Plymouth, or Dodge dealers in the U.S. The transmission and differential were made by Clark Equipment Company, and nearly all were the same as used in Dodge heavy-duty trucks. The Co-op was distributed through local cooperatives affiliated with National Cooperatives. Major wholesalers were Consumers Cooperative Association, Kansas City, Missouri; Central Cooperative Wholesale, Superior, Wisconsin; and Farm Bureau Oil Company, Indianapolis, Indiana.

Ford Motor Company, Detroit, Michigan

Although Henry Ford was already ambitious to build a tractor by 1907, the inimitable Fordson did not appear until 1916. It was an instant success, and launched an entirely new era in the tractor business. During 1918 alone, over 34,000 Fordson tractors were built and sold. Production continued apace until ending in 1928. Despite its faults, and despite the criticisms leveled against the Fordson, it was highly instrumental in mechanizing American farms. After the Fordson arrived, American agriculture would never be the same; the Fordson quickly became an icon as well as a role model.

The Fordson tractor was, for the 1917 market, the ultimate in simplicity. All moving parts were enclosed. The tractor was equipped with Ford's special air washer; it heralded the beginning of air cleaners as standard equipment throughout the tractor industry. In addition, the Fordson had the advantage of ready access to repair parts through some 22,500 Ford car dealers. This alone was a distinct advantage over having to order repair parts from a distant company and waiting for several precious days until they arrived. Few companies could match Ford on price, and this probably

After domestic production of the Fordson ended in 1928, the company concentrated on the European market, as well as importing these tractors into the United States. This continued through the 1930s. The English Fordson was a vastly improved version of the original style, although it retained many of the original features. Electric starting and lighting became optionally available, and the 1938 Fordson Allaround was Ford's first attempt at a row-crop tractor.

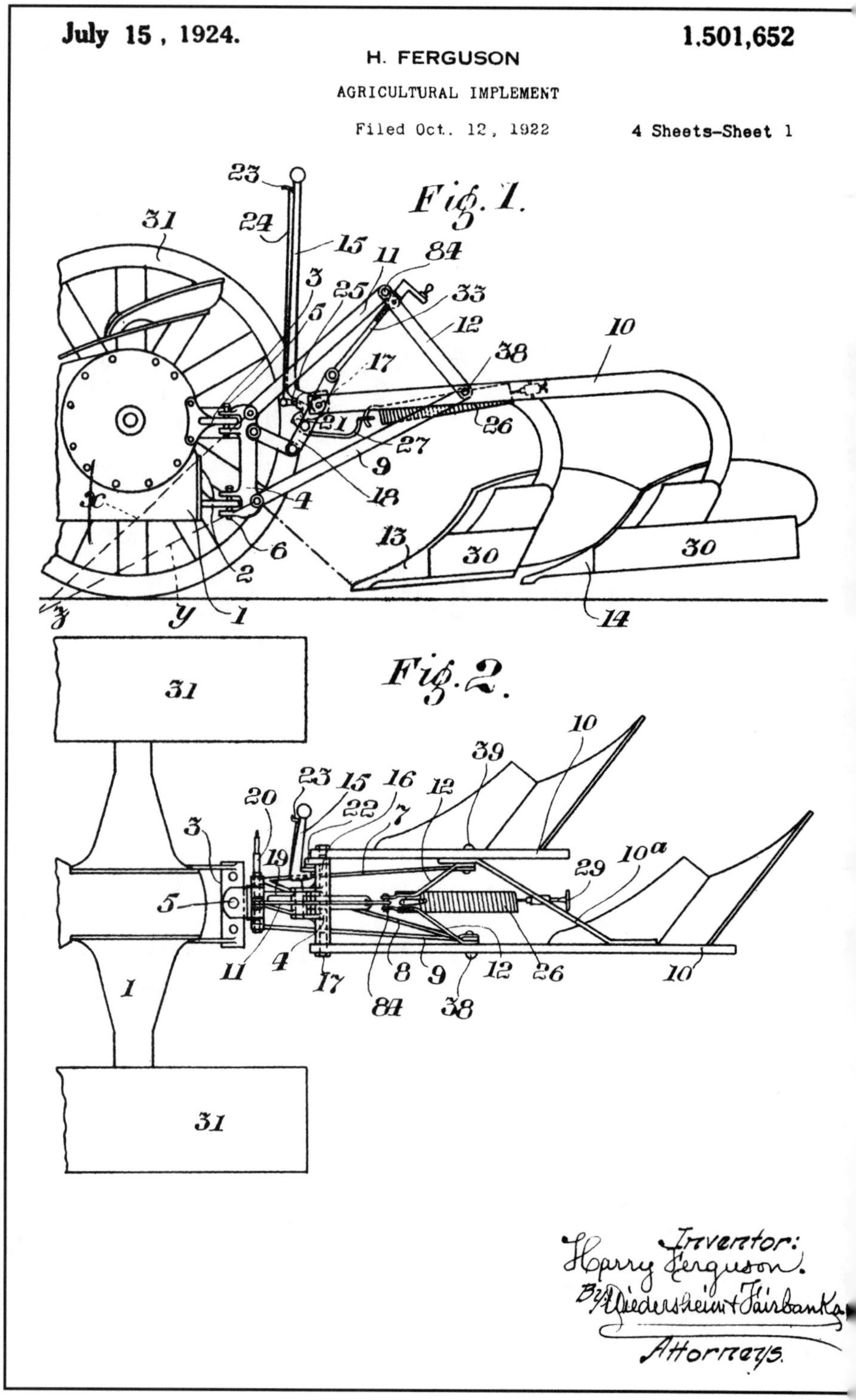

Shown here is one of the patent drawings for the three-point hitch system developed by Harry Ferguson. Although this concept made slow headway for several years, Ferguson convinced Henry Ford of its value. The result was the Ford 9N tractor, which first appeared in 1939. This was the first production tractor to use the three-point hitch system, and eventually, Ferguson's three-point hitch would once again revolutionize the design of farm tractors worldwide.

Four Drive Tractor Company, Big Rapids, Michigan

By 1916 the Fitch FWD tractor was being marketed, and this design remained viable into the 1930s. Numerous companies attempted to put traction onto all four wheels, with varying degrees of success. Due to the complex engineering problems involved, most of the four-wheel-drive tractors of this period were not entirely satisfactory. Because of its longevity in the tractor business, if for no other reason, the Fitch FWD apparently was among the best of the litter for this era.

Fox River Tractor Company, Appleton, Wisconsin

This company was incorporated in 1919, ostensibly to build tractors. One of its pre-production models is shown here, with the production model being very similar. Perhaps because of the fierce competition in the tractor market, or perhaps for other reasons, the Fox tractor was only on the market for a short time. Fortunately, the company turned its attention to building ensilage cutters, and in this regard they were eminently successful. Fox River Tractor Company became far more famous for its ensilage cutters and related equipment than for its tractors.

Franks Tractor Cultivator Company, Owensboro, Kentucky

By 1915 tractor development had moved to the point that farmers were clamoring for a machine that would cultivate standing crops. Thus, the second major phase of tractor development began. First came the plows that turned over the prairie with big, heavy tractors; now came a need to replace animal power with tractor power when cultivating the crops. One such response was the Franks Motor Cultivator of 1917. Like many other tractors of the time, this venture had a very short life span.

Geiser Mfg. Company,
Waynesboro, Pennsylvania

Geiser began building steam traction engines in 1881, and made its first venture into the gas tractor business about 1910. This one embodied numerous parts from its steam traction engines, including the kerosene headlight mounted on the front of the radiator. For reasons unknown, Geiser never made an impact in the tractor business, perhaps because it preferred to concentrate their efforts on an established line of grain threshers and other machinery.

General Motors Corporation, Pontiac, Michigan

In 1918 General Motors bought out the Samson Iron Works at Stockton, California. It moved the plant to Janesville, Wisconsin, and continued manufacturing the latter's Sieve-Grip tractor. This unusual design used an open-face drive wheel, thus the unique tradename. Although GM built these tractors for a year or so following the acquisition, it announced its own Samson Model M tractor in December 1918. It was of conventional design, and was intended to counter the domination of the Fordson tractor. This Sieve-Grip is shown pulling a nine-foot McCormick combined harvester.

About 1920 General Motors bought the manufacturing rights for the Jim Dandy Motor Cultivator. The result was the Samson Iron Horse. It was a line-drive tractor, meaning that with two leather reins or lines, it could be driven much like driving a living, breathing horse. Design flaws, market resistance, and fierce competition doomed the Iron Horse to a hasty end.

Gile Tractor & Engine Company, Ludington, Michigan

Gile began business in the early 1900s by making marine and stationary engines. By 1913 the company was building small tractors, and these evolved through several styles and sizes, even including a crawler model. The tractor shown here was one of its last models, being made about 1919. It was an early example of the unit frame design whereby the engine block, clutch housing, and transmission case were bolted together to form the backbone of the tractor. Gile was eventually bought out by Stearns Engine Company and disappeared from the tractor scene.

Graham-Paige Motors Corporation, Detroit, Michigan

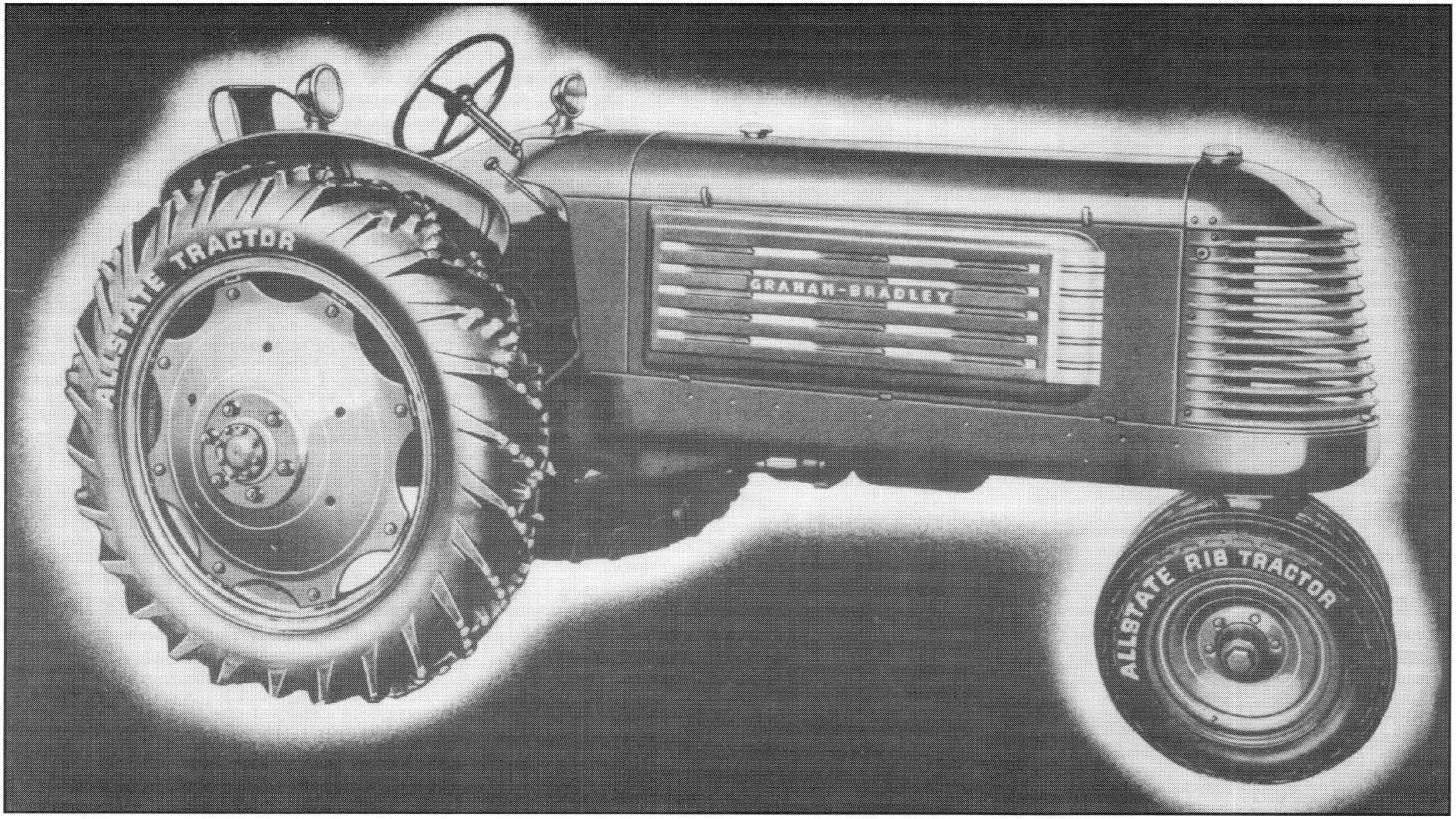

The Graham-Bradley tractor appeared in 1938. This model featured a fully stylized hood and grille, being as attractive as anything on the 1938 market. Graham-Paige was in the automobile business at the time, and its six-cylinder car engine powered the tractor. The Graham-Bradley was sold by Sears, Roebuck & Company for a time; thus the Allstate-brand tires on

Gray Tractor Company, Minneapolis, Minnesota

The Gray tractor was in development as early as 1908, but the model shown here first went on the market in 1914. Despite its unique design, the Gray Drum-Drive tractor remained on the market for a number of years, with production finally ending in 1933. The large drive drum at the rear precluded the need for a differential, it minimized soil compaction, and provided excellent flotation. An automatic plow guide is attached to the front axle in this illustration.

Hackney Mfg. Company, St. Paul, Minnesota

In 1911 the Hackney people came out with their version of the ideal tractor. The operator was furnished with an upholstered seat, much like an expensive touring car. Ordinarily a two- or three-bottom plow was mounted beneath the tractor, but as shown here, a road grader could also be attached. In this situation, two people were needed. Of course, the tractor operator was in the usual position, but the grader operator stood on the footboard at the rear of the tractor to raise, lower, and otherwise control the

Hart-Parr Company, Charles City, Iowa

In 1901 Hart-Parr Company built its first tractor at Charles City, Iowa. The unique part of the story is that the company's new factory was the first one in history that was devoted to building tractors. Up to that time, other companies had some other machinery coming out the door, such as threshers or steam traction engines. Hart-Parr had none of these, opting instead to devote all of its energy to the tractor business. This tractor of thirty belt horsepower was the first of several models more or less experimental in nature, but it led directly to Hart-Parr No. 2 of 1902 and all the others that would follow.

By 1903 Hart-Parr was firmly established in the tractor business. Included were tractors like the 22-45, a two-cylinder model that was immensely popular. It was followed by the famous 30-60. Also known as "Old Reliable," this tractor was widely accepted during its 1907-18 production run. Hart-Parr was one of the first manufacturers to make extensive use of steel castings, along with hardened steel gears. The concept was to lighten the tractor, but all the while increasing the strength of the parts.

The early Hart-Parr tractors made use of a unique planetary gearing arrangement; it is clearly visible within the belt pulley. Massive gearing was an integral part of these early tractors, but this was necessary to withstand the heavy loads to which they were subjected. Eventually, engineering technology and changes in design would obviate the need for these heavyweight components.

In the 1908-14 period, Hart-Parr built a 40-80 tractor, one of the largest tractors on the market. It used a massive four-cylinder opposed engine, with a pair of cylinders on each side of the crankshaft. Two separate radiators are obvious to the rear of the tractor, as are the gasoline, water, and kerosene tanks to the rear. The operator had a commanding view after ascending the ladder and getting situated at the controls.

Very little information can be found concerning the Hart-Parr 60-100 tractor. Built in the 1911-12 period, it isn't even known how many copies were made, or even any of its basic dimensions. This gigantic tractor was the largest ever made by any American manufacturer, and rumors have circulated for many years that a few of them made it to Montana or another of the western states. This huge tractor is said to have weighed in excess of 26 tons.

In 1918 Hart-Parr pioneered an entirely new tractor line with its New Hart-Parr, later known as the Hart-Parr 30. As noted on the front of this tractor, Hart-Parr proudly pointed to its pioneering efforts, with the words, "Founders of the Tractor Industry" being emblazoned on the front casting of this Hart-Parr 30. Hart-Parr claimed to have coined the word "tractor" as an acronym for the conventional term of "gasoline traction engine." Although the word itself is very old, it was Hart-Parr who made effective use of the word for its own machines, and it was Hart-Parr who popularized the word so that it eventually became part of the vocabulary.

In the 1914-16 period, Hart-Parr, like many other manufacturers, succumbed to the then-current trend of building a three-wheel design. Its Hart-Parr Little Devil, built during that period, was extremely simple, but never became popular. Although the company hoped to capitalize on the fad of the day, farmers were reluctant to accept a little tractor with a two-cycle engine, and a design altogether different than the Hart-Parr tractors with which they were familiar. Thus, after a short foray with this design, production ended, and the company once again concentrated its efforts on conventional designs.

Hart-Parr 30 tractors were offered from 1918 to 1922. They were a successor to the new Hart-Parr 12-25 developed the previous year. The Hart-Parr 30 was an immediate sensation, and was a strong competitor with the IHC Titan, and others of its horsepower class. Although the final drives were open gears, the remainder of the gearing was fully enclosed. The simple design permitted easy servicing and repair, and the automotive steering was a major improvement over the chain-and-bolster steering of the past.

By 1924 the Hart-Parr line was again ready for change, and the small 12-24 tractor appeared; it was the successor to the 10-20 that made its debut in 1921. The basic design of all the Hart-Parr models was the same, although minor changes occurred during the production run. A major improvement was the use of a plate clutch, as shown on this 12-24, in place of the band clutch previously used.

Various style changes occurred during the production run of the 12-24 and other Hart-Parr models. The old Hart-Parr 30 eventually was replaced by the 18-36, and as these changes occurred, various design changes were also incorporated into the new tractors. Toward the end of production, in the 1928-30 period, Hart-Parr once again modified its design. Even though the tractor was essentially the same as before, various small improvements helped keep it salable in a very competitive market.

In 1923 Hart-Parr introduced its 22-40 tractor. This one used a four-cylinder engine that was essentially a pair of 10-20 engines on a common crankshaft. When it was phased out in 1927, its replacement, the 28-50, was essentially a pair of 12-24 engines mounted side by side. The 28-50 was a massive tractor and was the largest model of the line. Production of the 28-50 ended in 1930.

In the late 1920s, somewhere between 1928 and 1930, Hart-Parr announced an optional power take-off shaft attachment for the 18-36 tractor. As can be noted in this photograph, it was driven from the crankshaft counterweight with a roller chain, crossed beneath the frame, and after going through a transfer box, power was delivered to the rear of the tractor. Since this device was driven directly from the crankshaft, a clutch was necessary to control the power shaft. In effect then, this was the first live PTO system to be offered on a farm tractor, and was somewhat ahead of those in the 1940s who claimed to be the first to offer this device.

J.A. Hockett, Sterling, Kansas

In 1893 the Hockett tractor appeared, it being the design of J.A. Hockett. Very little is known of this tractor, except that after two or three years the company ran out of money, and the enterprise was abandoned. Of significance, though, is the fact that this was the first gasoline traction engine to be advertised in any of the farm or trade magazines, even though a few others had built tractors prior to 1893.

Hollis Tractor Company, Pittsburgh, Pennsylvania

The Hollis 15-25 tractor began production in 1917 and continued into the early 1920s. This unique design used two close-set drive wheels which were also steerable. With this design the Hollis could actually make a square turn, and that was unusual for a tractor of 1917 vintage. As noted many times in this book, the range of tractor designs in the 1910-30 period is sometimes beyond comprehension.

Stockton, California

Holt was the first company to build a practical and commercially successful crawler tractor, doing so in 1904. The company merged with C.L. Best in 1925 to form Caterpillar Tractor Company. This photograph from the 1920s illustrates what it was like in the "good old days" as a Holt drags logs from the woods. Mud, brush, and tree roots are a small challenge for this tractor, even though the total weight of the tractor has been boosted by several hundred pounds of sticky muck.

Homestead Tractor Corporation, New York, New York

While most tractor efforts up to the 1930s were directed toward the farm tractor, numerous companies were actively mechanizing home and commercial gardens. These included the Auto-Tiller that came onto the market prior to 1920. The tractor shown here was heavy, weighing over half a ton, but it was capable of plowing and cultivating small plots which heretofore had been tilled by hand or by a horse and stirring plow.

Huber Mfg. Company, Marion, Ohio

Huber attempted to get into the tractor business in 1898. This model used a gigantic single-cylinder vertical engine. At this point, the methods of atomizing fuel were not well understood, with the vaporizer having potential to solve the problem. This engine is equipped with a vaporizer; with this device, fuel was maintained at a constant level in the small tank to the left of the engine cylinder. The intake air passed over the liquid fuel, drawing its evaporating vapors into the cylinder. This device was quite unsatisfactory, so probably for this as well as other reasons Huber left the tractor business and didn't return until 1911.

In 1911 Huber introduced its "Farmer's Tractor." It was a small machine that evolved into the 1912 model shown here. Heavy pintle-and-roller chains were used for the final drives, and the engine was of two-cylinder opposed design. Belt power was controlled by a clutch pulley visible in this illustration; it was engaged and disengaged by a hand wheel in the center of the pulley.

Huber was a well established manufacturer of grain threshers and steam traction engines. The company originated in 1865, and grew continually in the ensuing years. During 1912 Huber began building a huge 30-60 tractor as shown here. This massive tractor used a four cylinder engine. A unique feature was that each cylinder exhausted separately instead of the usual common exhaust through a manifold.

A side view of the gigantic Huber 30-60 tractor illustrates the massive bull gears and exceptionally large drive wheels. Partially visible are the four separate exhaust stacks from the big four-cylinder engine. The operators platform appears almost as an afterthought, but at least it was provided with a safety railing. Production of this tractor was apparently quite limited.

In 1917 Huber abandoned its heavyweight tractors in favor of smaller models using a crossmounted engine. Numerous sizes and styles were available, with the crossmounted styles remaining on the market for about ten years. Huber then offered a series of conventional unit frame tractors until all tractor production was suspended during World War II. Beginning in 1937 the company also built at least two different row-crop designs.

Huron Tractor Company, Detroit, Michigan

In 1916 or 1917 the Boyer Four tractor appeared for a very short time. In fact, a single advertisement in a 1917 issue of *Farm Implement News* magazine is the only reference this writer has found on the Boyer Four. The design includes automotive-type steering, full fenders, and attractive hood lines. Also featured was an upholstered seat, complete with backrest. Further details of the Boyer Four are presently unknown.

Illinois Tractor Company, Bloomington, Illinois

In 1916 this firm was known as Illinois Silo Company, reflecting its business interests up to that time. However, in that year the company introduced a small traction chassis to which a customer could mount his own gas engine and instantly have his own motor cultivator. Sales were apparently substantial enough to warrant further models, and in 1919 the company introduced a rather attractive conventional design. After 1921 no further information on the company can be located.

International Harvester Company, Chicago, Illinois

Of all the tractor manufacturers, International Harvester built more sizes and models than anyone. Of all the farm equipment manufacturers of the day, IH was the first to enter the tractor business, doing so in 1906. A friction drive system was used on the earliest models, with this engaged in a hill climbing demonstration in 1906. This tractor had a rating of 15 belt horsepower.

Another view of the 1906 International Harvester tractor illustrates the friction drive system. A fiber pulley on the engine crankshaft could be put in contact with the huge flat pulley located between the engine and the rear driver in this photograph. This system was effective and simple; both of these elements were necessary for a successful design. The full canopy was in keeping with the customs of the day.

The IHC Titan tractor line was built at Milwaukee Works, established in 1908. The IHC Mogul line came from Chicago Works, built in 1910. The former was sold by Deering dealers; the latter by McCormick dealers. Due to antitrust problems stemming from the large 1902 merger IHC was forced to operate several separate entities under its corporate umbrella for several years. The big Mogul 30-60 shown here is an early offering from Chicago

International Harvester built its 30-60 Mogul in the 1911-17 period. Although this tractor exemplified the best in design, its popularity paled against some of its competitors. However, due to the multitude of sizes and models offered by IH, this company led the industry in sales during 1910 and 1911. Here's a couple of 30-60 tractors turning over sod and converting open prairie into productive crop land.

I. H. C. TITAN 30-60 H. P. THROTTLING GOVERNOR KEROSENE TRACTOR

While the Mogul was being built at Chicago Works, the competing, and entirely different 30-60 Titan tractor was coming from Milwaukee Works. This tractor was of entirely different design than the Mogul, but fully met the manufacturing standards that had made the company famous. Sales of this tractor ran from 1910 to 1917. Originally marketed as the Titan 45, this tractor eventually was given a 30-60 horsepower rating; the latter more accurately reflected the tractor's capabilities.

Big tractors like the Titan 45 required lots of muscle power for cranking. To obviate this problem, the 1913 model shown here was equipped with an air starting device. A small IHC engine-air compressor combination was mounted at the front of the tractor. Once the air reserve was brought up, starting the engine was relatively easy with this equipment. The Titan 45 was one of the very few, if not the only, farm tractor to be regularly furnished with air starting equipment.

By 1912 International saw the need for a tractor smaller than the 30-60, and began building its 15-30 tractor. It was better suited to the average farmer, even though it was still a rather large, cumbersome, and expensive tractor. Although sales of the 15-30 were fairly good, the company was already working on tractors that were small, efficient, and reasonably priced. Production of the 15-30 continued until about 1918.

Production of the Mogul 8-16 tractor began in 1914. This single-cylinder tractor soon became tremendously popular, with the company sometimes calling it "The King of Small-Farm Tractors." The 8-16 was exceedingly simple; it was relatively easy to operate, service and repair, and it was relatively inexpensive. In 1916 the 8-16 was slightly remodeled and re-rated as the 10-20 Mogul, with production of the latter ending in 1919. The Mogul tractors and engines were sold by McCormick farm equipment dealers.

While the Mogul tractors were sold by McCormick dealers, the Titan tractors were sold by Deering farm implement dealers. The Titan 10-20 tractor came onto the market in 1915 and was an immediate sensation. These two-cylinder tractors came onto thousands of farms during their production run, which ended in 1922. Over 17,000 were built during 1918 and 1919, followed by over 21,000 in 1921. Undoubtedly, the 10-20 Titan helped immensely in establishing International Harvester as a dominant force in the farm tractor industry.

The Titan 10-20 tractors aided greatly in the farm mechanization process. Their simplicity was an important feature. When used for belt work, the clutch pulley, visible in this photograph, was operated by the handwheel at the outer end of the shaft. For plowing a special plow guide attachment was available to minimize work for the operator. It attached to the shaft extension clearly shown in this photograph, and was raised and lowered with a rope convenient to the operator.

International 8-16 tractors first appeared in 1917. This small, compact tractor was the first of the IHC line to use a four-cylinder vertical engine and a cellular radiator. With various modifications, the 8-16 saw a production run lasting until 1922. Considering the extremely rapid developments in tractor design, five years was indeed a long time for any company to build the same tractor model.

Always on the cutting edge, International Harvester had developed a new unit frame design by 1920, and first put it on the market in 1921. This new tractor was the 15-30, and it became an icon in its time. It embodied everything that IHC engineers had to offer, and became very popular during its long production run. In 1929 the 15-30 was redesigned to become the New 15-30; it was also known as the 22-36 tractor. Proof of its popularity is the fact that over 57,000 units were built in the 1929-34 period, even when the U.S. was in the throes of the Great Depression.

Already in 1916, IHC was experimenting with various tractor designs in its search for the ultimate tractor. After countless prototypes, the Farmall tractor appeared in 1923; it was the world's first row-crop tractor design. By the end of the decade the Farmall was revolutionizing farm mechanization, with many other companies following suit and building their own row-crop models. Then in 1931, the Farmall was supplemented with the F-30 (Farmall-30), with the F-20 coming the next year. By the thousands, these all-purpose row-crop tractors came onto American farms and in fact, all over the world. The truth was that because of its international

An endless number of attachments were available for Farmall tractors, not only from International Harvester, but from other manufacturers as well. This rare photograph shows a Farmall tractor with a specially built Stover burr mill mounted to its drawbar. With this setup a farmer could take the equipment wherever needed, and still have the drawbar free for other farm duties. This photograph was taken at the Stover Engine Works, Freeport, Illinois.

With the advent of the Farmall row-crop tractor came a new determination to perfect the corn picker. Until the 1930s, most corn was picked by hand, with this task being one of the last major farm jobs to be mechanized. International Harvester pioneered much of the development which eventually yielded a successful corn picker, and in the 1930s it developed a two-row picker that mounted directly to the tractor.

The small F-12 tractor first appeared in 1932, with over 123,000 of them being built until production ended in 1938; it was then followed by the similar but slightly more powerful F-14 tractor for a short time. Here was a small row-crop tractor adaptable to small farms, and was also very suitable for the light jobs on larger operations. The entire F-Series tractor line was the basis for further developments, which led to the 1939 introduction of an entirely new and extremely modern tractor line.

In 1939 the new Farmall tractor line appeared. The F-Series tractors were now history, and in their place came new models that included the A, B, C, H, and M tractors. The latter was the largest of the row-crop models. Its fancy hood and radiator styling reflected the work of the best designers of the day, while the engine and drivetrain epitomized tractor engineering of the 1930s. This design remained intact until the early 1950s. At that point a new numbered series of tractors appeared, although they followed much of the design pioneered with the Farmall M shown here.

Actual production of the McCormick-Deering TracTracTor began in 1929, but experimental models were afloat for several years before that time. Many of the early experiments were based on attachments that replaced the rear wheels of the 10-20 and 15-30 tractors; aside from this, a few independent companies built crawler attachments for these tractor models. During the 1930s the IH crawler tractor line developed, including models like this T-40. By the 1940s IH was a major player in the crawler tractor industry, topping its line with the gigantic TD-24 in 1947. At the time, this was the largest crawler tractor on the market. For a variety of reasons, the IH crawler tractor line declined, and virtually ended in the 1960s.

Kansas City Hay Press Company, Kansas City, Missouri

The first K.C. Prairie Dog tractor appeared around 1916. It featured the well known Waukesha engine; this one was four cylinder with a 4 1/4 x 5 3/4 inch bore and stroke for approximately twelve drawbar horsepower. Only one drive wheel was used; this design was rather popular at the time. However, the overall design has an uncanny resemblance to the row crop tractors which were on the horizon.

Kenison Mfg. Company, Solomon, Kansas

Here's a tractor with a rather conventional appearance that appeared in 1918 or 1919. It was marketed for such a short time that there's virtually no historical information in the various guides and directories of the period. Rated at 12 drawbar and 24 belt horsepower, it was equipped with a familiar Waukesha engine. Although the Kenison might well have been a very good tractor, the withering competition at the time was probably a major factor in its early demise. Hundreds of other aspiring tractor manufacturers met the same fate.

Killen-Strait Mfg. Company, Appleton, Wisconsin

A few crude photos are the extent of information that has been located on this interesting Strait Endless Chain Tread Tractor. By the time this advertisement appeared in 1914, Holt Mfg. Co. at Stockton, California, had exclusive trademark rights to the term, "Caterpillar." Thus, other manufacturers of crawler tractors were forced to use other titles, such as "crawler" or "tracklayer" to describe their machines. Strait's tractor was only marketed for a few years.

Case 10-20 Gas Tractor

J.I. Case, Racine, Wisconsin

In 1915 Case introduced their 10-20 model, an interesting three-wheel design. The 10-20 was advertised as late as 1922.

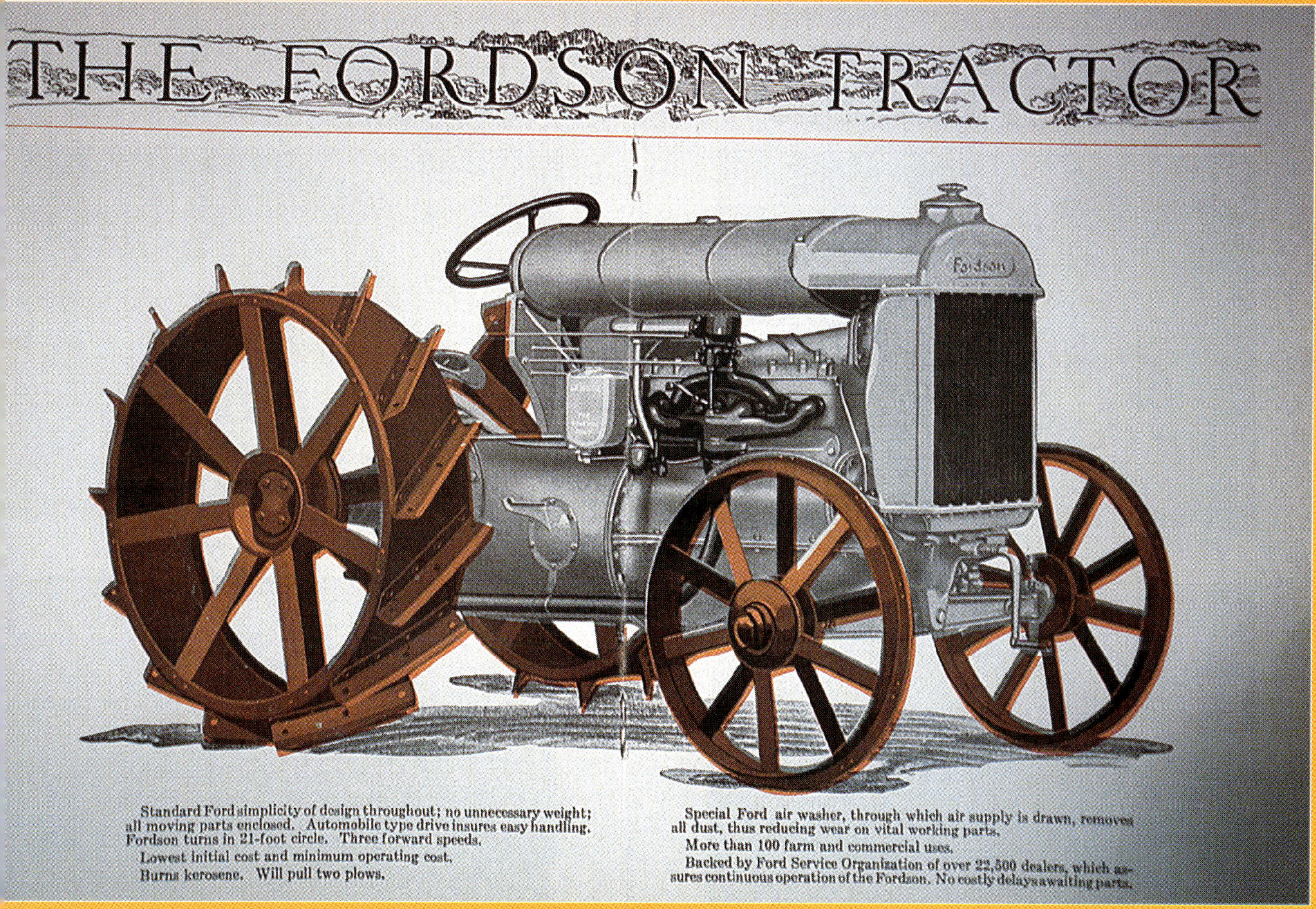

Ford Motor Company, Detroit, Michigan

One of the most popular tractors of all time, the Fordson was a simple design with no added frills. There were no fenders, and the belt pulley was extra, as was the governor.

Heider Mfg. Co., Carroll, Iowa

Heider Co. began building tractors in 1911, and this picture is of their first model. In 1914 Heider sold out to Rock Island Plow Co.

Deere & Company, Moline, Illinois

The Waterloo Boy was developed about 1913 by Waterloo Gasoline Engine Co. Deere made a buyout in 1918, and continued building the Waterloo Boy until 1923.

Graham-Paige Motors Corporation, Detroit, Michigan

The attractive Graham-Bradley tractor appeared in 1938. The company built tractors until 1946, at which time they merged with Kaiser-Frazer Motors to build automobiles. (Courtesy of the Chad Elmore Collection)

Oliver Corporation, Chicago, Illinois

ldition to the Oliver 70 Row-Crop, the company also offered an Oliver tandard. This fixed-tread alternative was preferred for certain applications, although the row-crop style predominated.

Emerson-Brantingham Co., Rockford, Illinois

Emerson-Brantingham emerged with this curious three-wheel design in 1916. Called the Model L, it was produced for only a short time.

Velie Motors Corporation, Moline, Illinois

Although well known in the carriage, automobile, and truck industries, the Velie tractor, built from 1916 to 1920, is virtually unknown among tractor collectors; none are known to still exist.

Ward Tractor Company, Lincoln, Nebraska

The Ward tractor operation was closely connected to Cushman Motor Works, and this 1912 model featured a Cushman 20 horsepower two-cylinder engine.

Minneapolis Threshing Machine Co., Minneapolis, Minnesota

In 1922 Minneapolis began building its 17-30 tractor, continuing with it until merging with other companies to form Minneapolis-Moline in 1929.

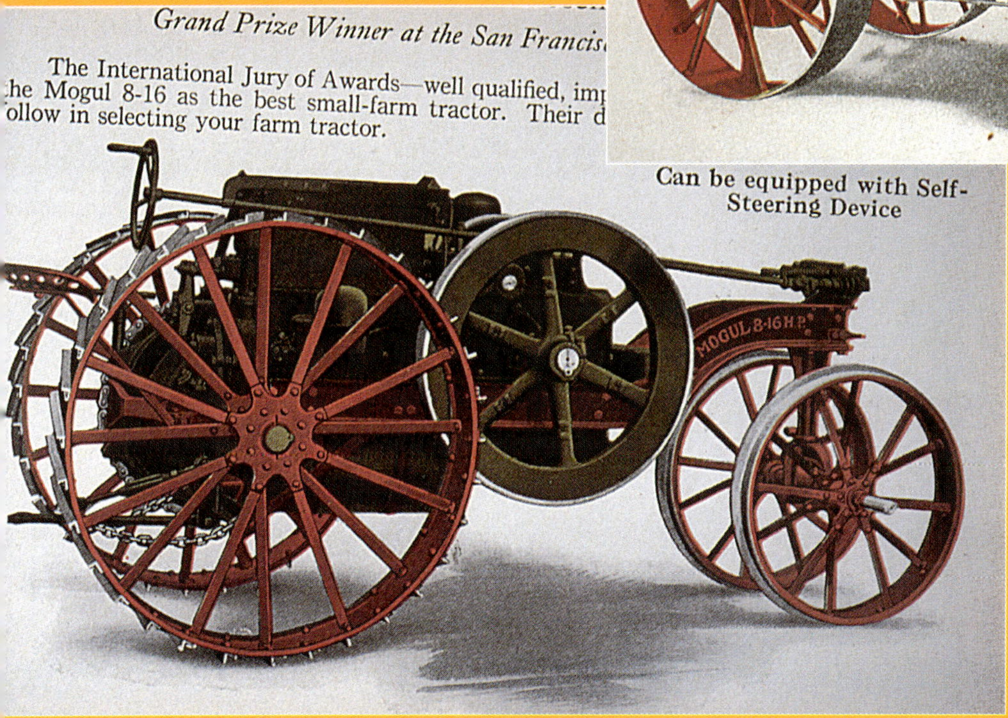

International Harvester, Co., Chicago, Illinois

The 8-16 Mogul from International Harvester was this comany's first entry into lightweight tractor designs, with the 8-16 gaining immediate popularity in the 1915 market.

R. H. Sheppard Co., Hanover, Pennsylvania

About 1950 the Sheppard diesel tractors entered the market. The Sheppard diesel engine was fully developed during World War II.

Minneapolis-Moline, Minneapolis, Minnesota

Minneapolis-Moline came out with its UDLX tractor in 1938. Only about 125 copies were made, with this being the first factory-equipped tractor cab.

Cockshutt Plow Co. Ltd.,
Brantford, Ontario

ockshutt line included an extensive series of tractor els, including the Cockshutt 50, an attractive design shed in red and yellow. This one is from the 1950s.

Illinois Tractor Co., Bloomington, Illinois

Illinois started making tractors about 1916, with this 18-30 tractor coming out in 1919, and remaining on the market until 1921.

John Lauson Mfg. Company,
New Holstein, Wisconsin

son was a big name in the gas engine business by the time this 1918 tractor appeared. The company remained in the tractor business for a decade.

J.I. Case Plow Works, Racine, Wisconsin

Wallis tractors, like numerous others of the 1920s, were finished gray, and trimmed with red wheels. For tractor enthusiasts the shade used on the Wallis is difficult to match and seems to range from a greenish gray to one with a bluish hue.

Shaw Mfg. Company, Galesburg, Kansas

During the Great Depression of the 1930s, conversion kits were an economical method of obtaining a small tractor. This one is a Shaw.

Minneapolis Threshing Machine Co., Minneapolis, Minnesota

Minneapolis 40-80 tractors first saw the market in 1912. This big tractor weighed about 11 1/2 tons and used a four-cylinder engine.

Allis-Chalmers, Milwaukee, Wisconsin

Allis-Chalmers began building their 18-30 tractor in 1919. For reasons unknown, this original company advertising shows the tractor finished in red. This curious anomaly has thus far found no plausible explanation.

Denning Motor Implement Company, Cedar Rapids, Iowa

Denning attempted to enter the tractor business about 1912, and was fairly successful for a few years. This color illustration is from the front cover of their catalog.

U.S. Tractor & Machinery Co., Menasha, Wisconsin

A 1920 catalog of the Uncle Sam Tractor included an artist's rendition of a tractor and plow on the front cover. Catalogs were a major sales tool of the 1920s, and this one was no exception.

Caterpillar Tractor Co., Peoria, Illinios

Caterpillar greatly expanded their crawler tractor line in the 1920s, with the Cat Ten and the Cat Fifteen being especially aimed at the business of farming.

Allis-Chalmers,
Milwaukee, Wiscon
An undated Allis-Chaln
catalog shows this 18-3
work pulling a plow

Allis-Chalmers,
Milwaukee, Wiscon
The Allis-Chalmers 20-35
a powerful and reliab
tractor that remained
production for many ye
It was essentially the san
the 18-30, which it repla
and was also followed b
Model E, 25-40 tracto

Avery Company, Peoria, Illinois
Tractor catalogs, such as this early Avery catalog, were major sales tools. Nearly flawless typographic design is evident, and this front cover epitomizes the printing arts of the early part of the century.

J.I. Case Plow Works, Racine, Wisconsin
The Wallis Model K emerged in the early 1920s as an evolvement of the Wallis Cub that originated in 1913. The Wallis was the first tractor to use a unit-frame design.

Rock Island Plow Co., Rock Island, Illinois

Heider tractors were very popular, especially in Iowa and Illinois. Their simple friction drive transmission was partially responsible for the success of this tractor. It was built into the 1920s.

Avery Company, Peoria, Illinois
During 1916 Avery came out with their first motor cultivator, and continued with various models for about four years.

very Company
eoria, Illinois.
ry came out with a
of tractors in 1912,
continued to aug-
and expand the line
next few years. All
tractors of the peri-
ollowed the same
ral design, varying
rily in physical size.

Avery Company, Peoria, Illinois
From the front cover of a 1911 magazine comes the only known color illustration of the first Avery tractor models. Sadly, none of these early Avery tractors are known to exist.

Cleveland Tractor Company, Cleveland, Ohio
A front cover design from a Cletrac catalog of about 1920 shows the tractor at work against an idyllic background. Cletrac was a strong competitor in the farm tractor business.

La Crosse Tractor Company, LaCrosse, Wisconsin

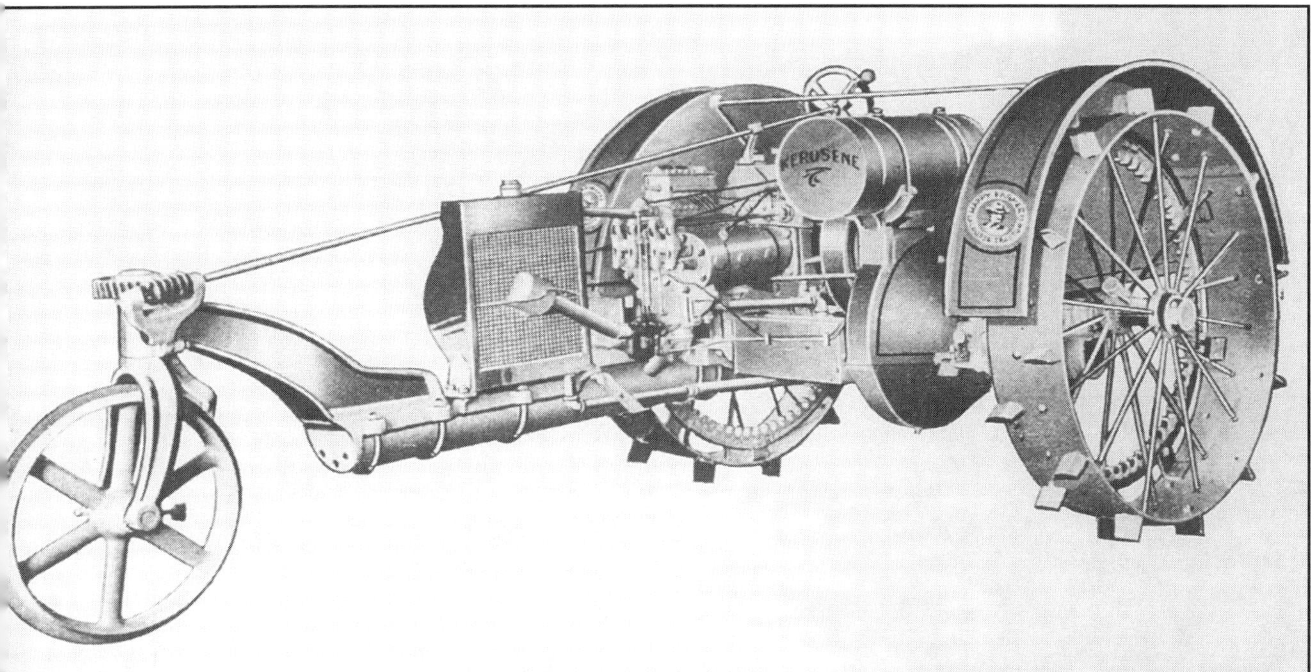

In 1916 the La Crosse Happy Farmer appeared, and made a considerable impression on Midwestern farmers. Various models were available, even including a line-drive tractor for those who were reluctant to part with their horses. At least with the line-drive model, the farmer could think of his tractor as a mechanical horse. Despite the sizable production of the Happy Farmer, the tractor industry was changing very rapidly, and this model was but a temporary means to an end. Ever-changing designs soon limited sales, and as a result, the Happy Farmer came to an end by 1922.

John Lauson Mfg. Company, New Holstein, Wisconsin

Lauson got into the tractor business in 1915, remaining there until the early 1930s. Lauson was one of the first tractor companies to manufacture a tractor made up primarily of OEM parts, sometimes referred to as an assembled tractor. By this time there were numerous engine manufacturers, and some of them specialized in tractor engines. Specialized companies were emerging in the transmission business, along with radiators, air cleaners, wheels, and other components. A distinctive feature of many Lauson tractors was the unique canopy or sunshade for the operator.

Lawter Tractor Company, Newcastle, Indiana

Between 1913 and 1916, Lawter built this unusual design with the plows under the center of the tractor. A major claim for this design was that it could plow a field right up to the fence corners. The company noted that with ordinary tractors, one-eighth of an acre was lost in each corner, while with a Lawter, even the corners of the field could be productive. Perhaps this argument was convincing, since the company remained in the market for three years, somewhat longer than many of its competitors.

Massey-Harris Company, Toronto, Ontario

After purchasing the J.I. Case Plow Works in 1928, Massey-Harris continued to build the Wallis tractors, albeit under the Massey-Harris trademark. The Model 25, also known as the 26-41 tractor, continued the same design that the Plow Works had pioneered with the Wallis Cub of 1913. However, this tractor of the 1930s was capable of over 41 belt horsepower using distillate fuel, and even more horsepower when burning gasoline.

Model 25 tractors were optionally available with pneumatic tires by the early 1930s. Late in the decade the design was streamlined to include a modicum of hood and radiator styling. Wallis tractors were painted gray, and some of the early Massey-Harris models appear to have been finished in dark green. Apparently, the company adopted its familiar red and yellow finish in the late 1930s, probably in conjunction with the announcement of the new stylized design.

The Pacemaker and Challenger tractors emerged from Massey-Harris in 1936. That same year the company also launched a short-lived venture with its four-wheel drive General Purpose model. The Challenger was a row-crop design, and was available with steel, as shown here, or with optional rubber tires. The original models were replaced with the Twin-Power design about 1938. As with all farm tractor manufacturers, production of farm tractors was virtually suspended from 1941 until 1945.

During 1938 and 1939 Massey-Harris abandoned the boiler plate frame design that had been used for a quarter century. In its place came an entirely new series of tractors using a cast iron unit frame with the engine mounted to the frame members. The 101 Standard shown here boasted a highly stylized design, but under the hood was a sturdy, powerful engine that quickly became very popular.

Model 81 and Model 101 Massey-Harris tractors were built primarily in the 1941-46 period, even though production was severely limited due to World War II. These new models were a great success for the company. In addition, Massey-Harris offered an extensive line of cultivators and other equipment that was tailor-made for these tractors.

In 1946 Massey-Harris launched an entirely new tractor line, with its Model 44 being the first to come off the assembly line. This new tractor, with a Massey-Harris four-cylinder engine, was also available in a standard-tread version, as well as special designs. Fuel choices ranged from gasoline to distillate to LP-gas; a diesel-powered version was also built. In addition, the Model 44 was the launching pad for an entire series of new tractors from the small one-plow Pony up to the massive Model 55. When the latter was announced, it had the highest horsepower rating of anything on the market.

In addition to the usual four-cylinder design, the Massey-Harris 44 was also available with a Continental six-cylinder engine. Some farmers preferred the smoothness offered by a six-cylinder engine, and this appears to have been the motivation on the part of M-H to offer this particular model. The 44-6 was built for ten years from 1947. In 1953 the Massey-Harris and Ferguson interests merged to form Massey-Harris-Ferguson. "Harris" was soon dropped from the corporate name, however.

Minneapolis-Moline Company, Minneapolis, Minnesota

Minneapolis-Moline was formed by the 1929 merger of Minneapolis Steel & Machinery Co., Minneapolis Threshing Machine Co., and the Moline Plow Co. Initially, the company continued to build many of the same tractors as its ancestors, but in 1931 its new Universal tractor was introduced under the M-M Twin City trademark. Weighing some 4,800 pounds, the Universal could be equipped with numerous mounted implements, including a variety of corn planters and cultivators.

After a few seasons the Universal was superseded by the Universal "J" tractor. This one had adjustable-width rear wheels, a front-mounted air cleaner and a redesigned steering system. It was also available with a "Toe-Tip" power lift system, and this feature was a major work saver for the farmer. Instead of raising and lowering implements with a hand lever, it was now a matter of pressing the button to do the same job.

In 1938 Minneapolis-Moline launched an entirely new tractor line. Highly stylized, it included a deluxe tractor in an open version as shown here. The concept was that a farmer could work in the field, and if a trip to town was necessary, there was nothing to it—except to put the tractor in road gear and make the trip.

Even though Minneapolis-Moline caught considerable press attention with its UDLX tractor of 1938, this fully enclosed model never sold more than about 125 copies. The UDLX even included a fancy chrome-plated front bumper, and chrome-plated hub caps on the rear axles. However, farmers of 1938 weren't ready for a tractor with an enclosed cab; twenty years later, the idea was gaining acceptance. (Courtesy of the Chad Elmore Collection)

Minneapolis-Moline offered an extensive tractor line in the 1940-69 period. The company pioneered the use of LP-gas as a tractor fuel. While other manufacturers were moving toward short-stroke, high-speed engines, M-M continued to build its long-stroke designs, and indeed, the M-M tractor had a sizable group of loyalists. Eventually, though, manufacturing farm tractors became a high-capital, high-risk business, and in 1969 Minneapolis-Moline, Oliver, and Cockshutt were merged into White Motor Corporation.

Minneapolis Steel & Machinery Company, Minneapolis, Minnesota

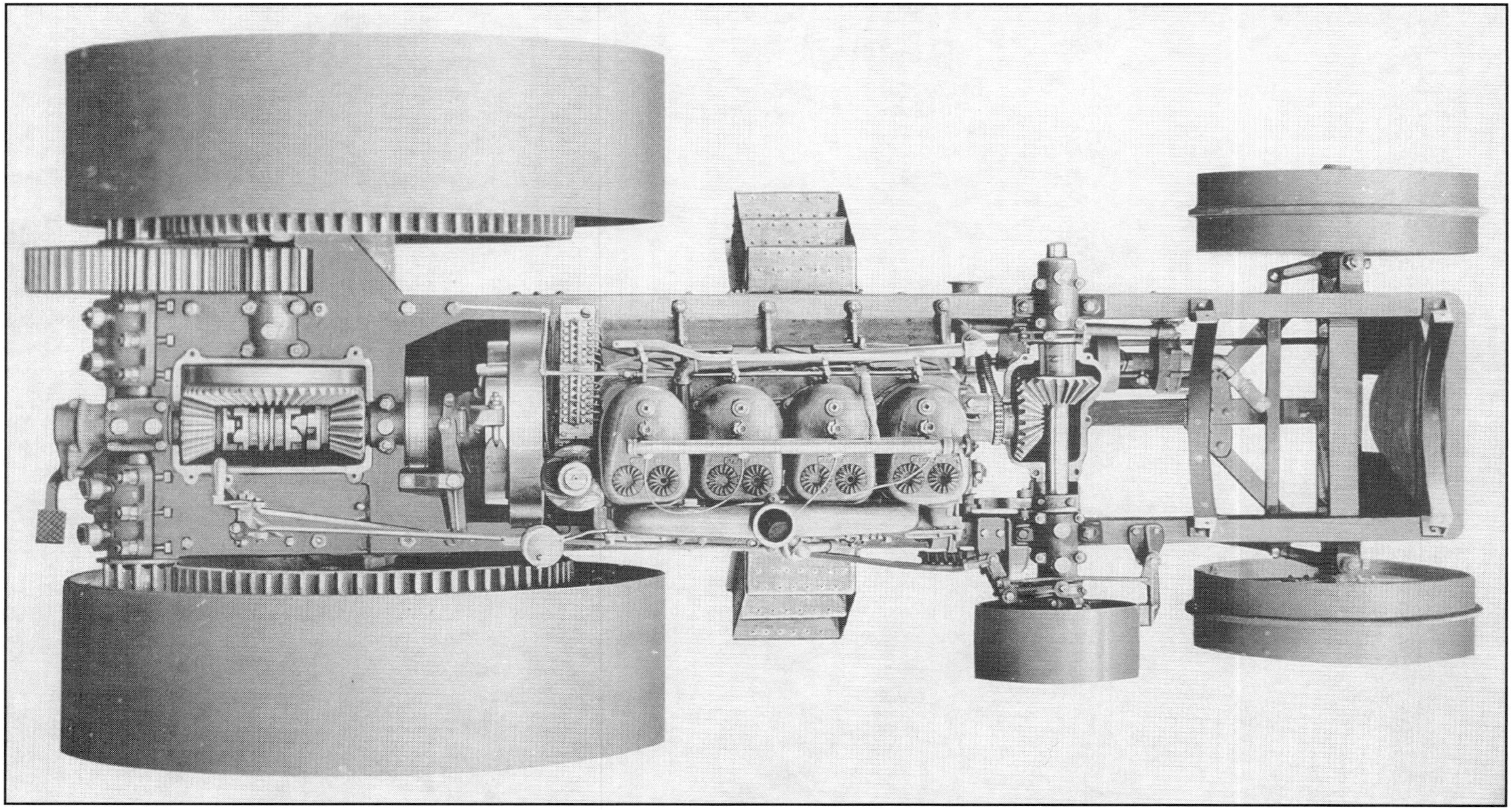

In 1910 Minneapolis Steel developed what would become its Twin City tractor line. The popular TC-40 is shown here in a top view for a better understanding of the design. The belt pulley was driven from the front end of the crankshaft, and jaw clutches are visible in the transmission case to effect forward and reverse travel. After first shifting to neutral, the engine was cranked from the rear

Beginning in 1913, Minneapolis Steel offered its huge 60-90 tractor. One of the largest tractors ever built, this one boasted a huge six-cylinder engine with a 6 1/4 x 8-inch bore and stroke. The 60-90 weighed fourteen tons and sold for $6,000. A substantial number of these tractors were sold for road building work. Undoubtedly, this huge engine consumed large amounts of gasoline every day.

In the 1913-17 period, Minneapolis Steel built three different versions of the TC-15 tractor. This small model carried a four-cylinder engine and utilized a cellular radiator. The design was essentially the same as in the larger Twin City models of the period, particularly the use of a massive steel frame to carry the engine and drive train.

The sporty 16-30 Twin City tractor first appeared in 1918. Fully hooded, it was an attractive design, especially for the 1918 tractor market. All gearing was enclosed; although the bull gears were merely shielded, at least they were kept out of the constant dirt and grit that severely shortened their life. Even the steering gear is enclosed and running in oil. This model disappeared after 1920.

An entirely new unit frame design was announced by Minneapolis Steel in 1919. The T-C 12-20 retained this rating until 1927 when it became known as the 17-28 tractor. This was the first tractor to use two intake and two exhaust valves per cylinder. Using this design obviously increased the power level over conventional practice, and for the 12-20 two separate camshafts operated the valve mechanisms. Minneapolis Steel, Minneapolis Threshing Machine Co., and Moline Plow Co. merged in 1929 to form Minneapolis-Moline.

Minneapolis Threshing Machine Company, Minneapolis, Minnesota

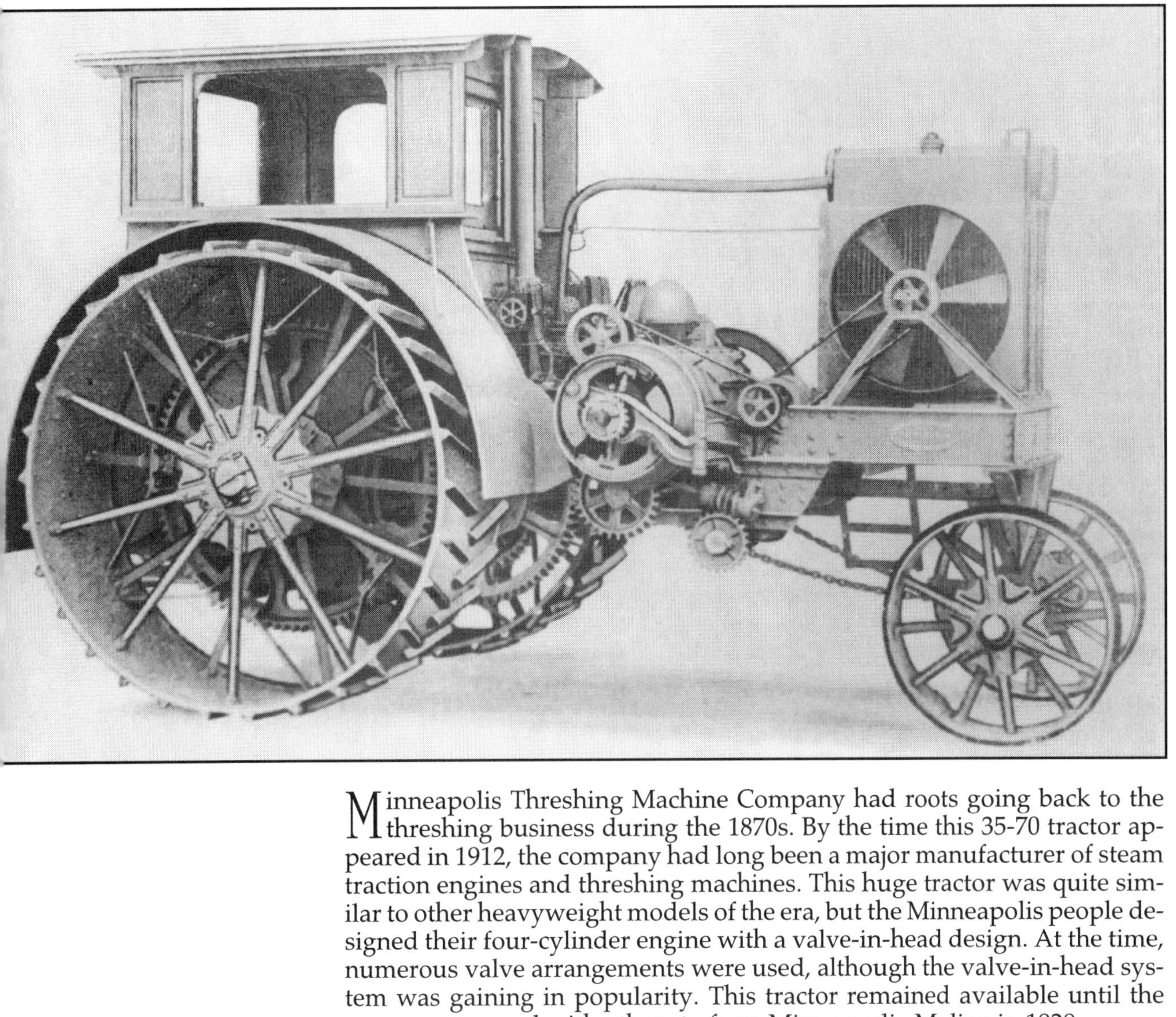

Minneapolis Threshing Machine Company had roots going back to the threshing business during the 1870s. By the time this 35-70 tractor appeared in 1912, the company had long been a major manufacturer of steam traction engines and threshing machines. This huge tractor was quite similar to other heavyweight models of the era, but the Minneapolis people designed their four-cylinder engine with a valve-in-head design. At the time, numerous valve arrangements were used, although the valve-in-head system was gaining in popularity. This tractor remained available until the company merged with others to form Minneapolis-Moline in 1929.

Minneapolis 35-70 tractors were originally rated as 40-80 tractors. However, when the tractor went through Nebraska Test No. 15 in 1920, the test results mandated a revised horsepower rating. In these tests the tractor shown here yielded over seventy belt horsepower. On the drawbar it pulled within fractions of 11,000 pounds, or 5 1/2 tons.

Late in 1915, Minneapolis Threshing Machine Co. introduced its version of a "small" tractor. Originally rated as a 15-30, it was rated downward to 12-25 in 1920, presumably because of its performance in the Nebraska Tractor Tests. However, this tractor remained in production until 1926. The 12-25 used a channel steel frame but had the distinct advantage of all gears being enclosed and running in oil.

In 1922 MTM built its first unit-frame tractor, the 17-30. It was built in two versions, with the 17-30 Type A having a four-cylinder engine of 4 3/4 x 7 inch bore and stroke. The later 17-30 Type B used 4 7/8 x 7 inch cylinder dimensions and had a slightly longer wheelbase; otherwise the two tractors were the same. Production of the 17-30 continued into the 1929 merger that formed Minneapolis-Moline Company.

Moline Plow Company,
Moline, Illinois

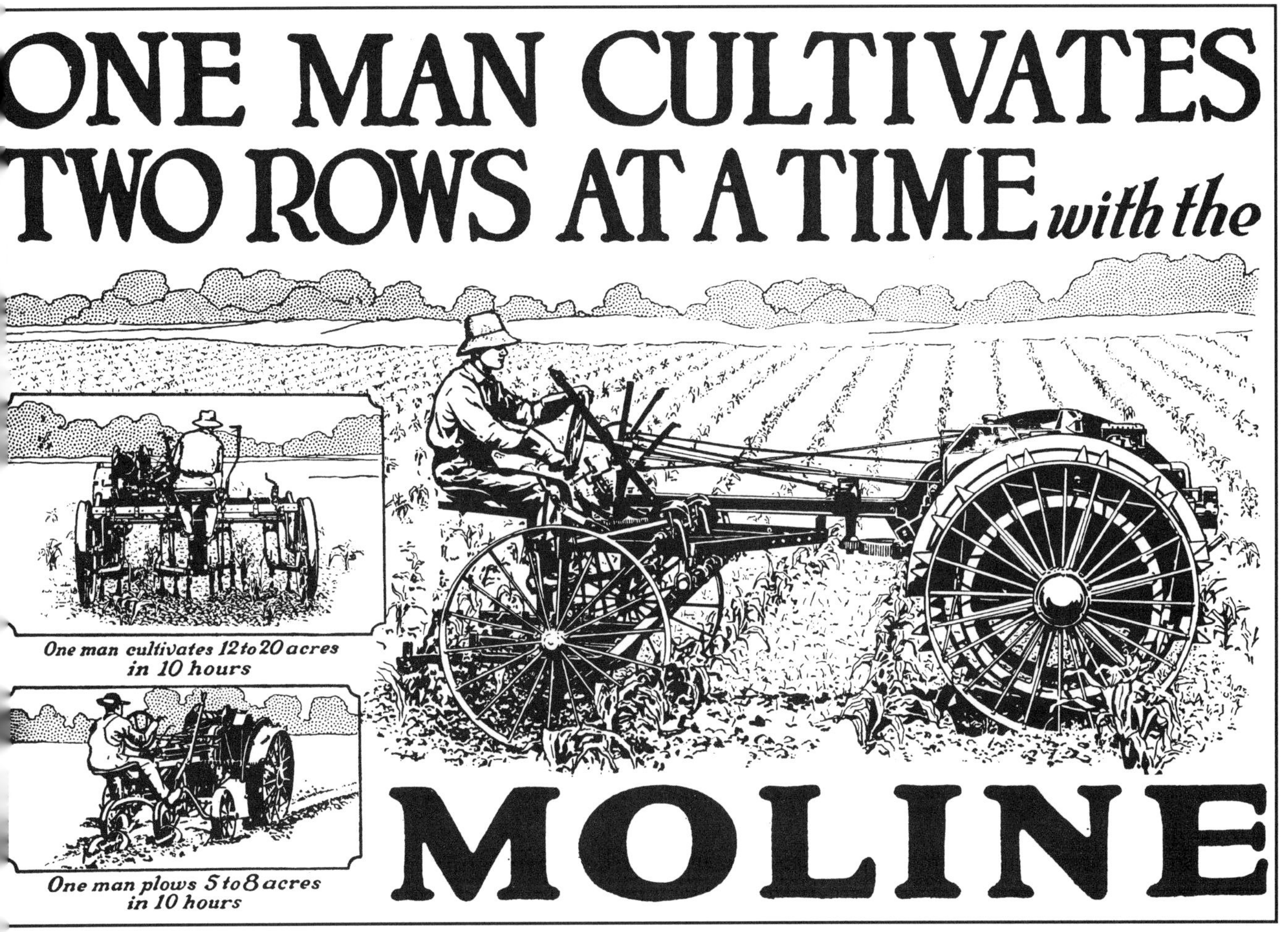

This company dated back to the 1850s as a small firm making farm implements. When the company bought out Universal Tractor Co. in 1915, it had by then become one of the leading farm implement manufacturers. From 1915 through 1917, Moline Plow built the Universal shown here; it was of two-cylinder design, but was quite capable of plowing, cultivating, and attending to other farm duties.

In 1918 the Moline Universal was redesigned to include a four-cylinder engine. The new model was equipped with electric starter and lights, and offered radically new features such as an electric governor. The company offered many different implements for use with the Universal, such as the plow shown here. In addition, the Universal could be adapted to grain binders, rakes, mowers, and many other farm implements. Despite its unique design, farmer enthusiasm for the Universal began to wane by 1920.

Production of the Moline Universal ended in 1923. Shortly before that time the company attempted to enter the construction market with a road scraper, as shown here. It does not appear that this unit got past the prototype stages. In earlier years, a road grader attachment was available for the Moline Universal tractor. After production ended in 1923 the company did not build tractors, but concentrated on its extensive implement line.

National Tractor Company, Cedar Rapids, Iowa

The National was the immediate successor of the Denning farm tractor. This firm was a subsidiary of General Ordnance Corporation of New York City. When Denning went bankrupt in 1919, National was the name first given to the tractor, but in December of that year the parent company renamed the tractor as the G-O, an abbreviation of General Ordnance. Regardless of the changes in name, the latter also was out of the tractor business in a short time, apparently by the end of 1920.

Nelson Blower & Furnace Company, Boston, Massachusetts

The concept of four-wheel-drive tractors gained limited popularity early in the tractor industry. Numerous companies attempted to build a four-wheel-drive tractor, but with varying degrees of success. Introduced in 1920, the Nelson tractor remained on the market for but a brief time. It was built in three different sizes.

Nichols & Shepard Company, Battle Creek, Michigan

Nichols & Shepard came out with its first tractors in 1911, and kept building tractors into the late 1920s. From the beginning, N & S used an extremely heavy design; even this 25-50 model of the 1920s is built up over a massive steel framework. While the company was well known for its popular Red River thresher line, its tractors were lesser known. Over the years the N&S tractors sold rather poorly in comparison to others in their class, and this despite the heavy, rugged design. However, the company was of singular mind, and preferred to continue offering these tractors to interested buyers. By the late 1920s the company offered the Lauson tractors as an auxiliary line, apparently up to the 1929 merger that formed Oliver Farm Equipment Company.

Nilson Tractor Company, Minneapolis, Minnesota

Nilson started business in 1913; its original design used three wheels. Two front steering wheels were used, along with a center-mounted rear drive wheel. Various models were built under this design, with the company finally adopting a four-wheel, standard-tread design in 1918. Two models were built, the Nilson Senior 24-36, and the Nilson Junior 15-30. At various times during a production run ending in 1929, the latter was also known as the Nilson Junior 16-25. This one is at work grading an unimproved road, known in the vernacular as a "mud road."

Nilson's original tractor design used a front axle mounted solid to the frame, with a single rear drive wheel. The operator sat in the center of the tractor, a position very advantageous for front visibility, but detrimental for keeping an eye on the towed implements. The single drive wheel eliminated the need for a differential mechanism and greatly simplified the drivetrain.

Ohio Tractor Mfg. Company, Marion, Ohio

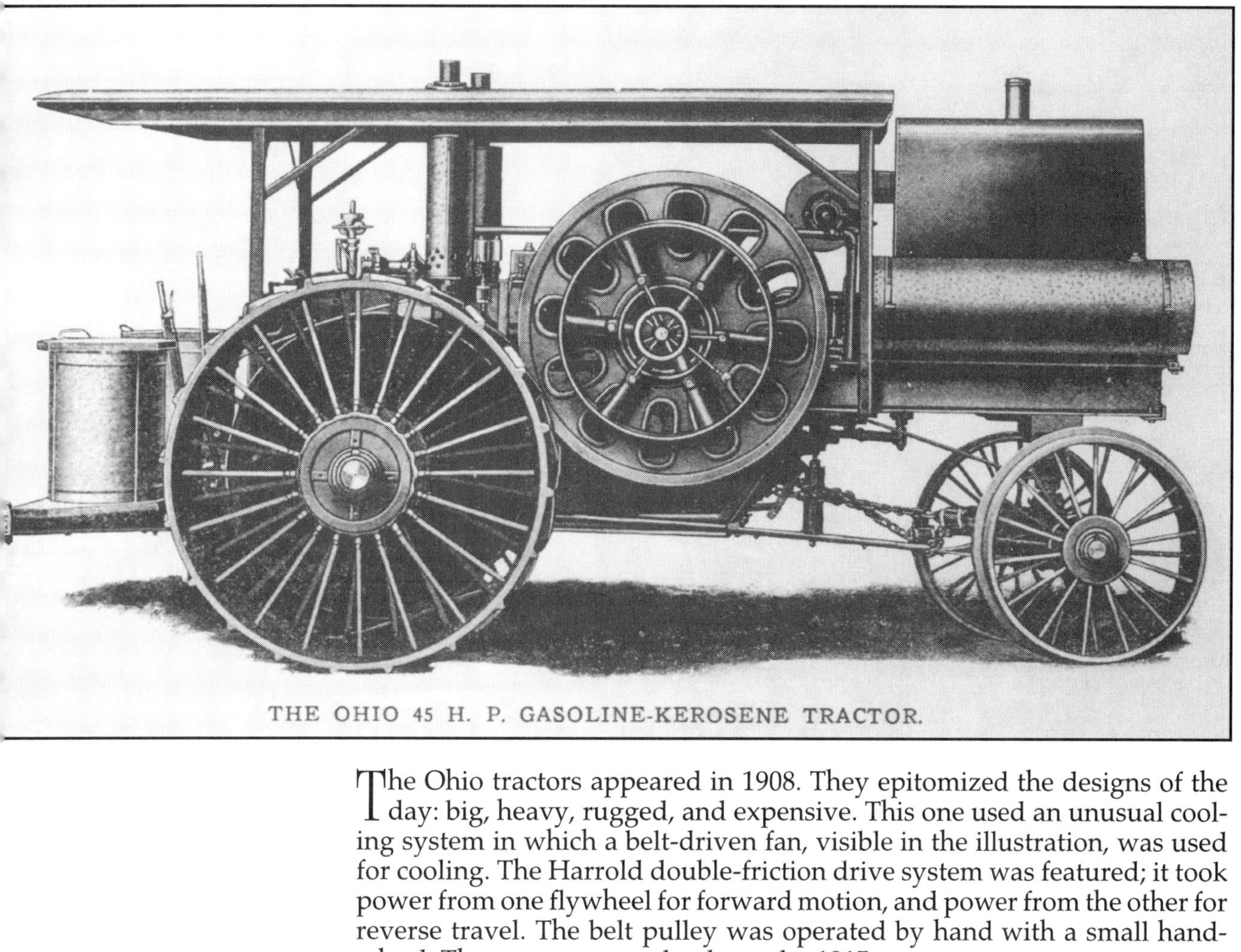

THE OHIO 45 H. P. GASOLINE-KEROSENE TRACTOR.

The Ohio tractors appeared in 1908. They epitomized the designs of the day: big, heavy, rugged, and expensive. This one used an unusual cooling system in which a belt-driven fan, visible in the illustration, was used for cooling. The Harrold double-friction drive system was featured; it took power from one flywheel for forward motion, and power from the other for reverse travel. The belt pulley was operated by hand with a small handwheel. The company was bankrupt by 1915.

Oliver Corporation, Chicago, Illinois

Oliver was formed by the 1929 merger of Hart-Parr Co., Nichols & Shepard Co., American Seeding Machine Co., and Oliver Chilled Plow Works. Immediately, the newly formed company set about designing new tractors at the Charles City, Iowa, plant of Hart-Parr Co. The result was the Oliver 18-27 Row-Crop in 1930. It was the forerunner of an extensive line. Building on the reputation gained by Hart-Parr the Oliver tractor line gained immediate acceptance.

In 1935 Oliver introduced its famous Oliver 70. Versions of this tractor were built until 1948. The six-cylinder engine provided operating smoothness seldom equalled among farm tractors, and the sleek styling was equally impressive. Meanwhile, the company was also building various other row-crop and standard-tread designs, but the Oliver 70 was the standard-bearer for the Oliver tractor line.

The Oliver 80 Row-Crop first appeared in 1937, and was built until 1948. In 1940 the company also launched its Oliver Row-Crop 80 Diesel tractor. Ultimately, less than a hundred were sold, but this tractor marked the company's entry into diesel-powered tractors. The 80 Diesel was equipped with a Buda-Lanova engine, developed by the Buda Engine Company at Harvey, Illinois. The latter firm was an engine manufacturer of worldwide fame, and eventually sold out to Allis-Chalmers.

In 1948 Oliver began building an entirely new tractor line, all the while phasing out the Oliver 60, 70, 80, 90, and other models developed primarily in the 1930s. In their place came the 66, 77, and 88 Row-Crop models. In addition, these tractors were also available in a standard-tread version as shown here with this Oliver Standard 66 Diesel model. The standard-tread models didn't see the production numbers enjoyed by the row-crop tractors, but were nevertheless an important part of the overall tractor line.

In the 1954-58 period, Oliver built its "Super" series tractors as an upgrade from the earlier tractor series of 1948 and following. Row-crop tractors were available in many different configurations to suit individual needs. Standard-tread tractors were likewise available, and any of these could be furnished for various fuels, ranging from gasoline to diesel or propane. During the 1950s Oliver experimented heavily with new gasoline engine designs, but by that time the diesel engine was gaining in popularity. In 1960 Oliver became a part of White Motor Corporation.

Parrett Tractor Company, Chicago, Illinois

In 1915 Parrett set up shop in Chicago to build its new tractor. It featured a crossmounted vertical engine, and for a few years at least, it was a sensation within the industry. Despite its popularity, the fierce competition among tractor manufacturers led to the company's downfall, and in 1921 it was reorganized as Hicks-Parrett Tractor Company, but this venture does not appear in any of the industry listings after 1922.

Peoria Tractor Company, Peoria, Illinois

Considering the short life of many tractor manufacturers, Peoria fared surprisingly well. By 1914 the company was offering a three-wheel design similar to the famous Bull tractor. The company built several interim models, finally coming out with this modern-looking 12-25 tractor in 1920. It remained on the market for three or four years. Despite its open final drive gears, the 12-25 Peoria featured smooth hood and fender lines and an auto-steer front axle.

Pioneer Tractor Mfg. Company, Winona, Minnesota

In 1917 Pioneer announced this interesting design. The rear drive wheels were set close together, with the final drive in the center. This precluded the need for a differential gear. Despite its rather unusual appearance, there are indications that the Pioneer Special was on the market for two or three years; it was then replaced with a conventional design. Pioneer began building tractors in 1909, but disappeared from the scene about 1927.

Port Huron Engine & Thresher Company, Port Huron, Michigan

Port Huron had its roots in the threshing machine business, going all the way back to the 1850s. In the 1880s the company began building steam traction engines. For a brief period, from 1918 to 1922, Port Huron was in the tractor business. Its 12-25 model featured a friction drive system for infinitely variable forward and reverse speeds, as well as an infinitely variable speed on the belt pulley. While the Port Huron is listed in the tractor directories for several years, few details of this tractor can be located.

Pullford Company, Quincy, Illinois

By 1918 the Pullford tractor attachment was available for converting an automobile into a tractor. As the brand name indicates, this attachment was designed especially for the Ford Model T. Numerous companies built tractor attachments, but the Pullford appears to have lasted longer than most; it was available into the 1930s.

During the 1930s, Pullford was offering two different models of its auto-tractor conversion kit. One was for the ubiquitous and nearly indestructible Ford Model T, while the other was for the Ford Model A; the latter having virtually the same qualities as the former. Although specific information is lacking, it appears that the Pullford attachments were available into the late 1930s.

Reliable Engine & Tractor Company, Portsmouth, Ohio

In the 1916-20 period Reliable appears to have built a substantial number of these tractors, although to the author's knowledge, few, if any, still exist. The Reliable 10-20 was also marketed by Fairbanks, Morse & Company in the 1916-20 period as its Fair-Mor tractor. This attractive little model used a two-cylinder engine having a 6 x 7-inch bore and stroke.

Rock Island Plow Company, Rock Island, Illinois

This major plow manufacturer got into the tractor business through a 1914 buyout of the Heider Mfg. Co., Carroll, Iowa. The Heider's unique feature was a friction drive system that obviated the need for transmission gears and a clutch. One of the big friction plates is obvious directly behind the belt pulley. This system was extremely simple; it also had the advantage of providing infinitely variable ground speed and pulley speed. It could also be shifted on-the-go, a concept that gained great attention in the 1960s.

Royal Motors Company, Napoleon, Ohio

In the early days of the tractor business the line-drive tractor gained slight popularity. The idea was that farmers had previously known no other method of driving *anything* except for horses or mules. With a line-drive tractor they could go forward, back, right, or left with a pair of leather lines in their hand; the tractor thus became a mechanical horse instead of an intimidating "tractor." The Royal was only marketed for a short time after its 1917 introduction.

Schofield Auto Tractor Company, Kansas City, Missouri

By 1920 there were well over fifty different companies making tractor attachments for automobiles. Some, like the Schofield shown here, could be driven on or off the tractor chassis at will. Chain drives and gearing provided the necessary speed reduction. Despite a brief popularity, most of the conversion kits were only moderately successful, primarily because automobile engines weren't built for the heavy power demands of tractor farming.

Shaw Mfg. Company, Galesburg, Kansas

Shaw built a variety of garden tractors and other farm equipment, beginning sometime in the 1920s. During the 1930s this tractor conversion kit became available. Known as the Shaw Tractorized Car, this one was intended to utilize Ford Model T, Model A, and Chevrolet cars no longer suitable for automobile service. Also, as shown in this illustration, it could be "driven by lines about as Old Dobbin was driven," obviously in a pitch to those still reluctant to part with four-footed horsepower as compared to four-wheeled tractor power.

R.H. Sheppard Company, Hanover, Pennsylvania

The Sheppard tractors first appeared in 1949, remaining on the market for a few years. Sheppard pioneered some innovative diesel engine developments, with the Sheppard diesel engines seeing substantial use in industrial applications. This advertising photo of about 1950 notes that "Sheppard Diesels were one of the very first to provide big double disc brakes, an independent power take off, an independent hydraulic system, a fully adjustable hydraulic seat, and a heavy-duty simplified Full Diesel engine."

An advertising folder of the 1950s illustrates a Farmall Model M tractor converted to Sheppard Diesel power. Sheppard claimed that the entire changeover could be made in less than a day. The company even provided heavy-duty starter brushes for the "M" starter, although a heavier 12-volt starter for the Sheppard Diesel conversion engine was optionally available. This writer has never seen or heard of the conversion shown here, outside of the advertising; whether any of these units exist, we simply do not know.

Siemon Tractor Corporation, Buffalo, New York

The Iron Horse tractor first appeared about 1919, but was built for only a short time. As otherwise noted in this book, the Iron Horse terminology probably had a palliative effect on farmers still resistant to the concept of tractor farming. The combination of the trade name with the line-drive feature hoped to obviate these objections. Company advertising also refers to this unit as the Siemon-McCloskey tractor, probably in reference to its designers.

Square Turn Tractor Company, Norfolk, Nebraska

In the 1915-17 period Square Turn attempted to enter the tractor market with this unique design. The drive wheels were mounted forward, with steering being controlled by a single rear wheel. As the name implies, this tractor was able to turn a virtually square corner, thus permitting the farmer to plow into the corners of a field. In fact, the company claimed this tractor was so simple that "any woman or boy can fill a man's place with a Square Turn tractor." Sales of this tractor were something less than might have been hoped for, and the company was finally sold at a sheriff's auction in 1925.

Standard-Detroit Tractor Company, Detroit, Michigan

Sometime prior to 1920 the Tracford tractor attachment appeared; it was listed into the early 1920s. This one was dubbed as "The Tractor Universal" and was built over a Ford Model T chassis. The Ford Model T cooling and lubricating systems were barely adequate for motor car service, much less for tractor duty. Thus, the conversion units like the Tracford were only suitable for light-duty farm work.

Standard Engine Company, Minneapolis, Minnesota

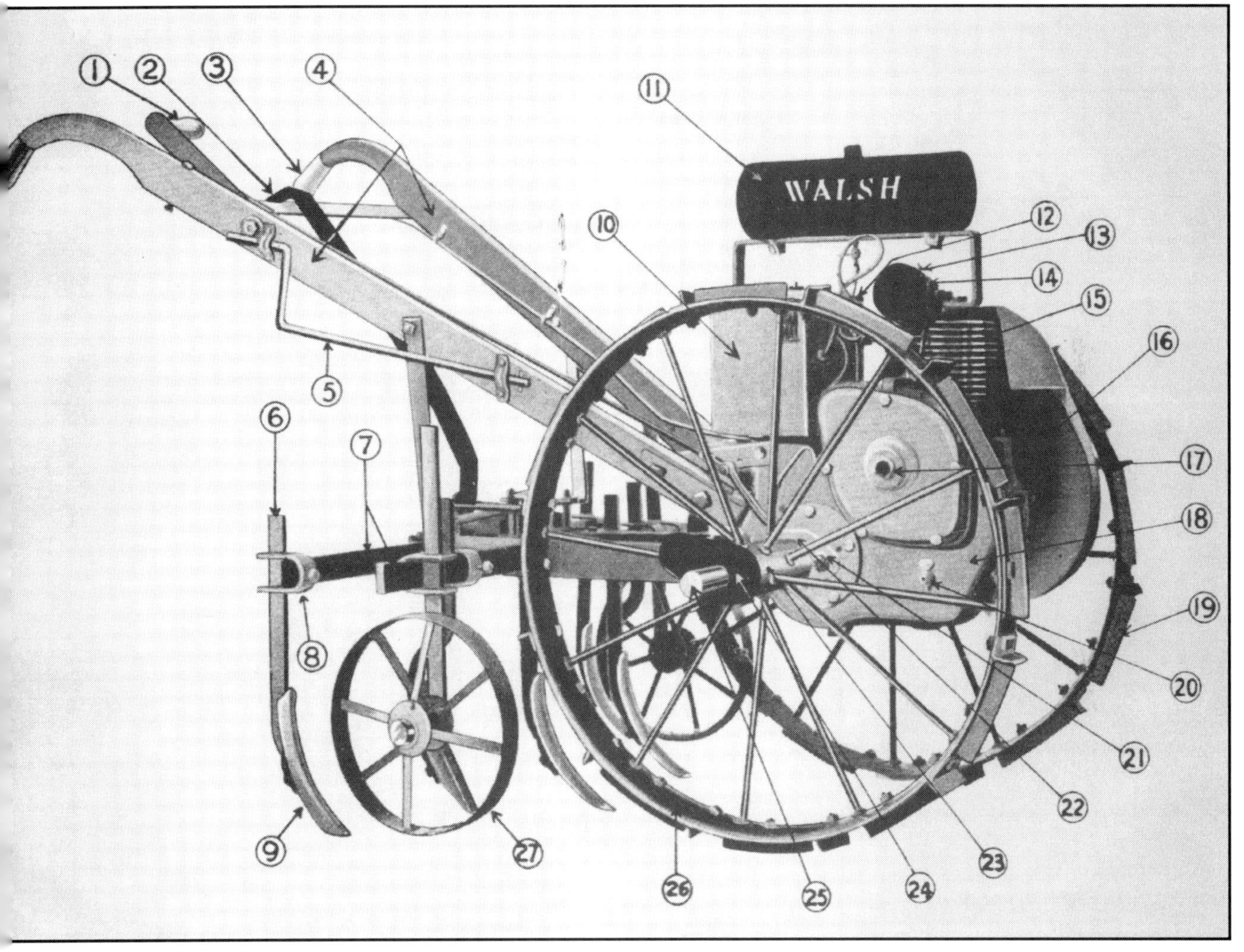

Walsh Tractor Company was formed by P.E. Walsh in the 1920s. Subsequently, the firm reorganized and became Standard Engine Company, with this Walsh garden tractor being one of its products. Although the Walsh garden tractors are well known among collectors, specific historical details on the company, its predecessors and successors, has been virtually impossible to obtain. The high-clearance design of the Walsh made it very popular among commercial gardeners.

Star Tractor Company, Findlay, Ohio

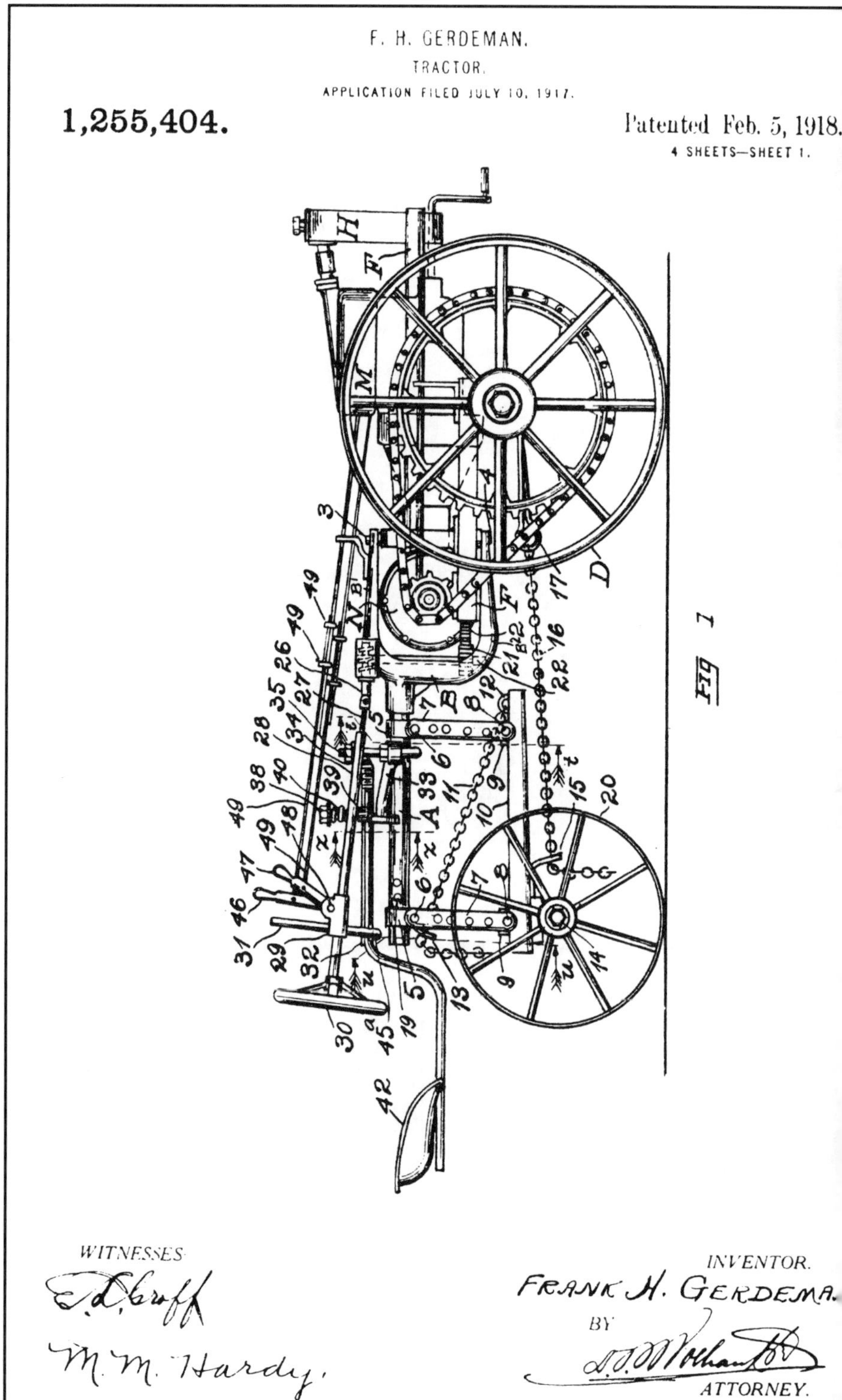

Outside of this patent, no information has been found concerning the Star tractor. It was a "universal" design, and when the patent was filed in 1912, this usually meant that the tractor was articulated, or hinged in the center. Numerous design features were included in the patent application, including an improved form of the joint where the tractor articulated. This is but one of many designs that never made an impact on the market, if indeed they got past the drawing board.

Stinson Tractor Company, Minneapolis, Minnesota

In the 1917-22 period, Stinson built tractors at Minneapolis, and after 1919, at Superior, Wisconsin. Although unconventional by contemporary standards, the design was quite in keeping with numerous tractor models of the time. A unique feature was the ability to move the front wheel to the right, to the left, or into the center, depending on the work to be done. For instance, when set to the right, the front wheel could then follow in the furrow when plowing.

A front view of the Stinson illustrates the simplicity of the design. Also illustrated is the long tiller lever by which the front wheel could be shifted to the right, to the left, or anywhere in between, depending on the job at hand. Stinson also featured totally enclosed gear cases with everything running in oil. For a design of 1917, this was indeed a major improvement.

In this view, a Stinson is shown with the front wheel running in the furrow while pulling a plow. While the unique chassis design was a handy feature, farmers of 1917 were being inundated with many other new designs, and with literally hundreds of tractor makes and models from which to choose, it was difficult for all but a very few manufacturers to dominate the market, or even to gain the slightest foothold. Thus, Stinson ended its career in the tractor business in 1922.

Toro Motor Company, Minneapolis, Minnesota

This company was organized about 1914, ostensibly to manufacture the Bull tractors. The name itself indicates the connection between Toro and Bull, since "toro" is the Spanish word for "bull." In 1923 the company announced this interesting motor cultivator that could be converted into a conventional four-wheel tractor. Production continued into the late 1920s. Advance-Rumely Thresher Company acquired the Toro line in 1927. After revamping it to the company's own ideas, it re-emerged as the Rumely DoAll tractor.

Townsend Mfg. Company, Janesville, Wisconsin

In 1914 Roy Townsend, a former engineer with Fairbanks, Morse & Company, began building the Townsend tractors. This unusual design looked very much like the steam traction engines it was intended to replace. However, what looked like the boiler was now the radiator, complete with tubes. The engine exhaust was piped vertically into the "smokestack," creating an induced draft through the "boiler flues" for cooling the engine. Built in several sizes, the Townsend was never sold in large numbers, disappearing completely from the market by the early 1930s.

Traction Motor Corporation, Muskegon, Michigan

This huge tractor was rated at 65 horsepower, and obviously included electric lights as part of the standard package. The huge operator's platform could likely have been equipped with a full cab as an option. Included only in the 1920 tractor directories, the Traction Motor was equipped with a big V-8 Spillman engine. It was one of the few tractors ever built with a V-8 engine until recent years.

Trenam Tractor Company, Stevens Point, Wisconsin

Trenam produced this interesting tractor design between 1917 and 1920. Rated at twelve drawbar horsepower, the Trenam followed the popular idea of a three-wheel tractor, placing the drive wheels forward, with a rear support wheel. An interesting feature is visible in this illustration. The arched frame to the rear of the tractor rested on the rear wheel support channels. When steering, the curved frame pivoted over the rear wheel support. The open shafts and drive chains would now be a safety engineer's nightmare.

U. S. Tractor & Machinery Co., Menasha, Wisconsin

In the 1918-22 period, this firm built the attractive Uncle Sam tractors. Two sizes were built, a 12-20 and the 20-30 model shown here. It was an extremely modern design for its time, with a fully hooded engine, plus full fenders for the convenience, safety, and comfort of the operator. In retrospect, the Uncle Sam may have goaded other manufacturers into streamlining their designs, but among today's vintage tractor collectors the Uncle Sam is merely another forgotten tractor.

Universal Tractor Mfg. Company, Columbus, Ohio

Universal was organized in 1914 to build this interesting motor cultivator. Priced at only $385, it was equipped with a two-cylinder, 10 horsepower engine. Late in 1915 Universal accepted Moline Plow Company's offer to pay $150,000 for the Universal Motor Cultivator. From this small machine came the famous Moline Universal tractor, the latter being a popular farm tractor for several years.

Victory Tractor Company, Greensburg, Indiana

Organized in 1918, Victory produced tractors at Greensburg until about 1921. As with many other tractors of this chassis style, it came close to being a row-crop tractor, but from all appearances, no one thought of carrying the design to this point. This 1919 model was of unit frame design. The unit frame, pioneered by the Wallis Cub in 1912, was quickly adapted by many tractor builders, and eventually became a standard in tractor design.

Ward Tractor Company, Lincoln, Nebraska

Ward Tractor Company was connected to the Cushman Engine Co., also of Lincoln, through L. M. Ward, the manager of Cushman Motor Works. This 1912 design was considered to be a lightweight tractor for its time, although it weighed some 6,500 pounds. Power came from a 20 horsepower, two-cylinder Cushman engine. The pintle-type final drive chains are evident, as is the induced draft cooling system termininating in a stack at the top of the water cooling tank. This popular and rather ingenious system terminated the exhaust in a nozzle above the cooling tank.

Wetmore Tractor Company, Sioux City, Iowa

The Wetmore tractor first appeared on the market around 1919, and remained there until the early 1930s. Although the Wetmore didn't achieve any high levels of production, it apparently displayed enough durability and reliability to keep it in the market, despite fierce competition. Rated at twelve drawbar and twenty-five belt horsepower, the Wetmore sold in the range of $1,300.

Wheat Tractor & Tiller Company, Buffalo, New York

The Wheat Tractor Ready for the Road.

The Wheat made its first appearance in 1917 under the auspices of Hession Tractor & Tiller Corporation, with the company name changing in 1921. The Wheat featured interchangeable steel wheels and hard rubber tires. Thus, the same tractor could work in the field, or on improved roads. By the 1920s, improved roads were much in demand for the newly developed automobile, and automobilists were so unhappy about steel-wheeled tractors tearing up the smooth roads that many counties and even some state legislatures prohibited "wheels with lugs" at this early date.

Whitney Tractor Company, Cleveland, Ohio

In 1918 the Whitney tractor appeared. This little tractor was rated at nine drawbar and 18 belt horsepower. The company was a successor to Ohio Mfg. Company; the latter had pioneered the Morton traction truck in 1902. Morton traction trucks were essentially a chassis to which an engine manufacturer could adapt their own engine and come up with a farm tractor. Production of the Whitney continued for only a short time, coming to an end sometime before 1920.

Wichita Tractor Company, Wichita, Kansas

During the 1918-20 period this small 8-16 Wichita tractor was marketed. The transmission and drivetrain are strikingly similar to the Waterloo Boy tractor being built at the same time. Even the front-mounted fuel tank bears a great similarity. One could easily ask why the Waterloo Boy was on the market for several years with sales numbering into the thousands, while this one was around for only a couple of years, with sales of only a few hundred likely. The answer is that no one really knows. Perhaps it was the color of the paint or the sound of the engine.

Willys-Overland Co., Toledo, Ohio

The inimitable Jeep was manufactured by the thousands for military use during World War II. Subsequently the company attempted to market the Jeep in the domestic market, even to the point of offering it as a versatile farm tractor. To this end, the Jeep could be furnished with belt pulley, power take off shaft, and a three-point hitch attachment. This underside view illustrates the core of the Jeep design.

Wolverine Tractor Company, Detroit, Michigan

Like hundreds of other companies, Wolverine entered the tractor market after World War I, ostensibly hoping for a share of the burgeoning tractor market. However, despite its rating of 15-30 horsepower, the Wolverine was over 12 1/2 feet. The long chassis precluded short turns, and this may have been detrimental to anticipated sales. After a very short time, this interesting tractor disappeared from the scene.

Yuba Mfg. Company, Marysville, California

In 1912 the Ball Tread Company at Detroit, Michigan, started building the Ball-Tread crawlers. Yuba bought out this firm in 1914 and continued building the Ball Tread design into the early 1930s. This very distinctive crawler design utilized a row of steel balls to carry the crawler tracks. While this design carried some obvious disadvantages, it completely eliminated the need for troublesome track rollers.

Bibliography

Numerous sources were used in the compilation of this book. Although the various books listed are currently in print, most of the magazines and directories, along with the original literature cited, are out of print.

Wendel, C.H. *Encyclopedia of American Farm Tractors*. Osceola, Wis.: Crestline/Motorbooks, 1979.

———. *150 Years of International Harvester.* Osceola, Wis.: Crestline/Motorbooks, 1981.

———. *American Gas Engines Since 1872*. Osceola, Wis.: Crestline/Motorbooks, 1983.

———. *Nebraska Tractor Tests Since 1920.* Osceola, Wis.: Crestline/Motorbooks, 1985.

———. *The Allis-Chalmers Story.* Osceola, Wis.: Crestline/Motorbooks, 1988.

Wendel, C.H. and Andrew Morland. *Minneapolis-Moline Tractors*. Osceola, Wis.: Motorbooks, 1990.

Wendel, C.H. *150 Years of J.I. Case*. Osceola, Wis.: Crestline/Motorbooks, 1991.

Wendel, C.H. and Andrew Morland. *Massey Tractors*. Osceola, Wis.: Motorbooks, 1992.

———. *Allis-Chalmers Tractors.* Osceola, Wis.: Motorbooks, 1992.

Wendel, C.H. *Oliver Hart-Parr.* Osceola, Wis.: Motorbooks, 1993.

———. *Fairbanks-Morse: 100 Years of Engine Technology, 1893-1993.* Lancaster, Pa.: Stemgas Publishing Co.

Magazines and Directories

American Thresherman magazine, various issues: Madison, Wis.

Chilton Tractor Journal, various issues: Philadelphia, Pennsylvania

Farm Implement News magazine: Chicago, Ill.
Farm Implement News Buyers Guide, various issues.
Tractor Field Book Annual Directory, various issues.

Tractor Red Book, various issues: Kansas City, Mo.: Implement & Tractor.

UNBEATABLE BOOKS FOR CAR LOVERS

GREAT GIFTS **GREAT READING**

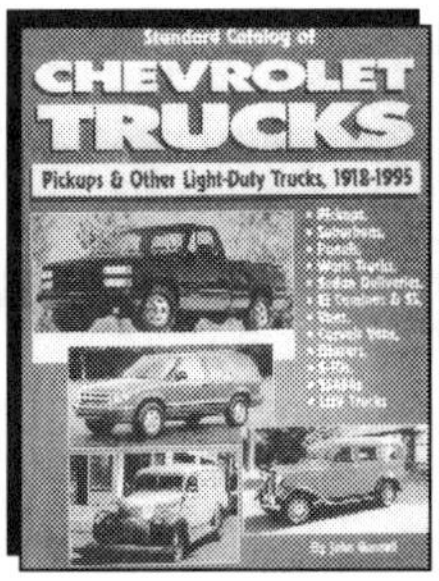

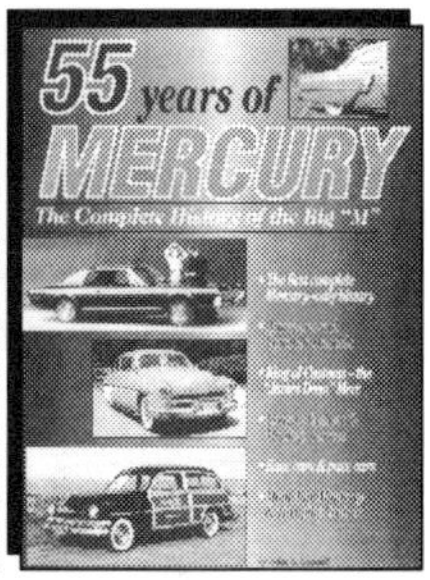

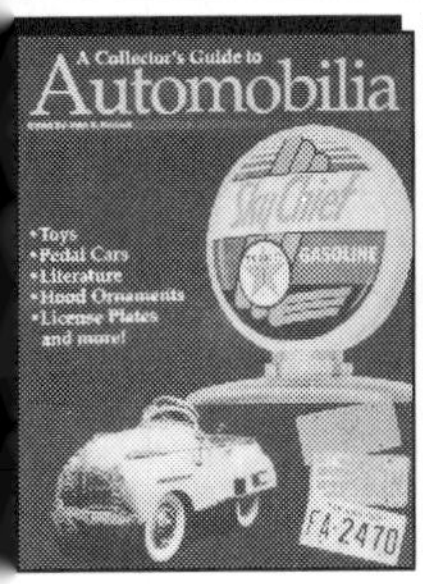

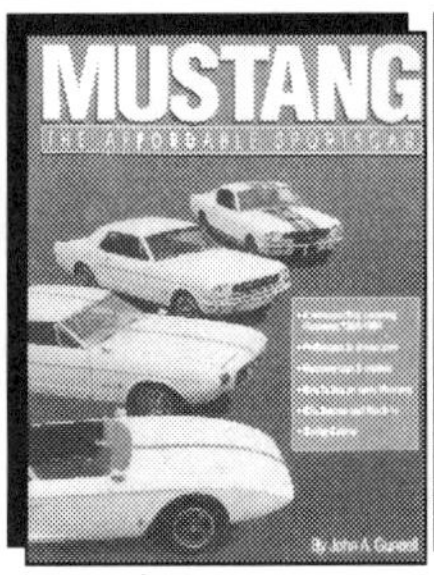

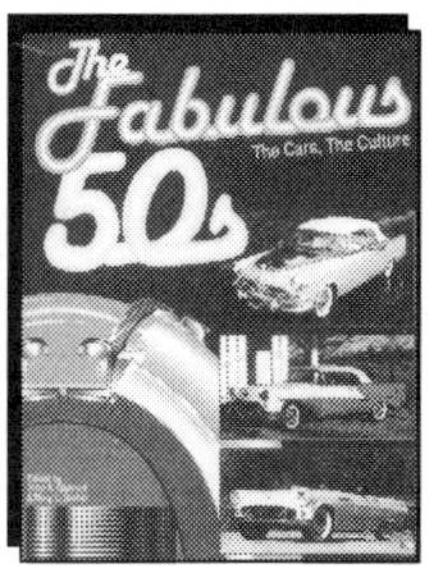

1996 STANDARD GUIDE TO CARS & PRICES, 8th Edition

Puts you in the driver's seat for the best deals on collector cars from 1901-1996. More than 150,000 current values give you the confidence to deal smart and provide you the marketplace edge.

6x9 SC • 656 p • 150 photos • CG08 $15.95

CORD 810/812, THE TIMELESS CLASSIC

Whisks you on a drive down memory lane in one of America's most recognizable cars. Enjoy hundreds of photos and drawings of one of the most dynamic autos of this century.

HC • 232 p • 200+ photos • CORD1 $34.95

CAR MEMORABILIA PRICE GUIDE

Is packed with values for memorabilia in up to six grades of preservation. Rare, clear photos help you identify and price even the most hard-to-find toys, models, hood ornaments, paper collectibles and more.

6x9 SC • 200 p • 400 photos • GAP01 $19.95

TRUCKING IN AMERICA - MOVING THE GOODS

Delivers a concise, informative and detailed photo history of how the highway haulers of yesterday forged the American economy of today. Hundreds of rare photos showing you all kinds of trucks from yesteryear.

SC • 304 p • 500 photos • ATR01 $16.95

WIRED FOR SUCCESS - AUTO ELECTRICAL MADE EASY

Helps you cut expense and do it yourself. With this step-by-step book, you'll discover the best tools and techniques for the electrical maintenance of your automobile.

SC • 240 p • 200 photos • AEM01 $19.95

STANDARD CATALOG OF PONTIAC 1926-1995

Provides technical specifications, fact-filled historical highlights, production totals, option lists and pricing in 6 grades. You'll also get expanded photo coverage in time to celebrate Pontiac's 70th anniversary.

SC • 304 p • 800 photos • APO01 $18.95

CARS OF THE STATE POLICE AND HIGHWAY PATROL

Arms you with an in-depth photographic look at the vehicles, past & present, used by America's state police agencies. Close-up photos of door markings and license plates are shown.

SC • 304 p • 500 photos • CSP01 $16.95

BETSY, BEST CAR I EVER OWNED, Auto-Biographies Of America's Favorite Cars

Captures your heart with a tribute to "Betsy" - an honored automobile portrayed in art and prose by Jeff Hagen. In short story form, owners tell why their "Betsy" won their heart.

6"x9" • SC • 180 p + 90 illus. • MY01 $14.95

PRESIDENTIAL CARS AND TRANSPORTATION

Illustrates the history of presidential transportation in this new book. Includes carriages, trains, yachts and planes. Takes you inside the cars of the '70s, '80s, and the '90s up to the current presidential fleet.

HC • 160 p • 100 photos • PCT01 $19.95

T-BIRD - 40 YEARS OF THUNDER

Touts the 40th anniversary of the luxury sport "T-Bird" and honors the car and the people who made it happen. You'll get complete details of this postwar classic from the early two-seaters to the mid-1960s four door Landaus.

SC • 304 p • 500 photos • ATH01 $24.95

STANDARD CATALOG OF CHEVROLET LIGHT-DUTY TRUCKS 1918-1995

Presents a treasure trove of information in this new addition to the series. You'll get technical specs, ID numbers, chassis specifications, engine data, and pricing in 6 grades.

SC • 304 p • 900 photos • ACT01 $24.95

55 YEARS OF MERCURY

Turns your spotlight on Mercury in this 55-year anniversary salute. "Big M" models appropriately named Turnpike Cruisers, Cougars, Comets, and Bobcats come to life in this historic look at Mercury, its makers and its market.

SC • 256 p • 250 photos • ME01 $16.95

AUTOMOBILIA - Automotive Memorabilia from 1900 to 1990

Shifts your collecting power into high gear! Find accurate values for collectibles, sources of factory photos, auto ad checklists, videos featuring classic cars and a historical survey of pedal cars

SC • 304 p • 300 photos • A001 $19.95

MODEL T FORD

Traces the evolution of the "Tin Lizzy" from 1908-1927. This book debunks popular myths, gives you data on specific Model T parts, lists engine serial numbers by day of production, and provides 1,367 photos presenting technical information.

HC • 600 p • 1,367 photos • TF01 $39.95

MUSTANG THE AFFORDABLE SPORTSCAR, A 30 Year Pony Ride

Presents a 30th anniversary tribute to America's favorite "pony." Loaded with over 300 photos, you'll read about Lee Iacocca's claim to fame, and why it still "is what it was"

SC • 304 p • 300+ photos • MS01 $16.95

FIRST HIGHWAYS OF AMERICA - A Pictorial History of Early Roads for Automobiles

Unravels the "how" and "why" of road building from 1890 through 1925. Join the salute of human willpower that wrestled, struggled, and persevered.

HC • 320 p • 200+ photos • HR01 $29.95

SENSATIONAL '60s

Details how the automobile helped shape America's history from 1960-1969 with more than 600 photos. Informative captions give you the facts, fads and philosophies of those turbulent years.

SC • 308 p • 500+ photos • SX01 $16.95

THE FABULOUS '50s - THE CARS, THE CULTURE

Takes you on a walk down memory lane with a photographic look at the cars that caught our fancy in the fabulous '50s. Over 500 photos, advertisements, and billboards are included.

SC • 304 p • 500 photos and 16 color photos • CF01 $14.95

Shipping and Handling: $3.25 first book; $2.00 each additional book.
Foreign Addresses: $10.00 first book; $5.00 each additional book.
WI residents add 5.5%, IL residents add 7.75% sales tax
All books are 8 1/2"x11" unless specified.
Dealers - call or write for more information. 715-445-2214

Send orders to: **KRAUSE PUBLICATIONS**
Book Department ZGB2
700 East State Street
Iola, WI 54990-0001

For faster service, credit card customers call **1-800-258-0929 Dept. ZGB2**

KRAUSE PUBLICATIONS FOR QUALITY, COLLECTING, HOBBY, OUTDOOR AND AGRICULTURE / TRADE PERIODICALS

From Antiques to Antlers to Automobiles... From Barbies to Buildings to Blades... From Cards to Coins to Comics... The ABCs of special interest periodicals for YOU

SPORTS

Krause Publications is the leading publisher in the sports collectible industry. Its periodicals range from Sports Collectors Digest, the hobby's oldest and largest publication, to Trade Fax, the late-breaking service for industry insiders, to Fantasy Baseball, the current craze for kids and adults.

Fantasy Baseball	6 issues annually	$11.95/yr.
Sports Card Magazine and Price Guide	Monthly	$24.95/yr.
Sports Collectors Digest	Weekly	$59.95/yr.
Trade Fax	Twice per week	$300.00/yr.
SCD's Price Guide Weekly	Weekly	$59.95/yr.

NUMISMATICS

These high-quality periodicals provide numismatists across the globe with comprehensive information on coins, paper money and memorabilia - the leading publications for collectors worldwide.

Bank Note Reporter	Monthly	$29.95/yr.
Coin Prices	Bi-Monthly	$16.95/yr.
Coins	Monthly	$24.98/yr.
Numismatic News	Weekly	$29.98/yr.
World Coin News	Monthly	$22.98/yr.

PHILATELY

Experts and novices, collectors and dealers, buyers and sellers are served with the most up-to-date publications available.

Stamp Collector	Weekly	$34.88/yr.
The Stamp Wholesaler	Bi-Weekly	$29.90/yr.

RECORDS/COMICS

Comics collectors, buyers and sellers are kept up-to-date and record enthusiasts can find their "Goldmine" through these high-quality, information-packed publications.

Comics Buyer's Guide	Weekly	$36.95/yr.
Comics Retailer	Monthly	$29.95/yr.
Goldmine	Bi-Weekly	$39.95/yr.

TOYS/ANTIQUES/COLLECTIBLES

Contemporary and limited-edition art, antiques, collectibles, show and auction listings - everything to capture the interest of buyers and sellers.

Collectors Mart	Bi-Monthly	$23.95/yr.
Today's Collector	Monthly	$21.95/yr.
Toy Shop	Bi-Weekly	$26.95/yr.

GUNS/KNIVES

The sharpest and most explosive publications in the field, targeting collectors, dealers, gunmakers, knifemakers and anyone interested in guns and knives with all the current information to meet their needs.

Blade	Monthly	$19.95/yr.
Gun List	Bi-Weekly	$29.95/yr.
Gun Show Calendar	Quarterly	$14.95/yr.

OUTDOORS

Leading edge hunting and fishing tactics, how-to methods, hunting and fishing adventure, biology and behavior combined with superb color photography for the hunting and fishing enthusiast.

Deer and Deer Hunting	8 issues annually	$17.95/yr.
Trapper & Predator Caller	10 issues annually	$16.95/yr.
Turkey & Turkey Hunting	6 issues annually	$12.95/yr.
Wisconsin Outdoor Journal	8 issues annually	$16.95/yr.

AUTOMOBILES

Up-to-the-minute references on pricing from the nation's most respected authority on value. Complete and quality coverage of antique and collectible automobiles.

Old Cars	Weekly	$34.95/yr.
Old Cars Price Guide	Bi-Monthly	$18.95/yr.

TRADE/AGRICULTURE

The entire spectrum of Wisconsin's agriculture industry, the nation's frame builders and the rural construction industry are covered in comprehensive fashion in these quality publications.

Agri-View	Weekly	$40.00/yr. ($28.00 WI)
Frame Building News	5 issues annually	$20.00/yr.
Rural Builder	7 issues annually	$18.95/yr.

kp since 1952

KRAUSE PUBLICATIONS

700 E. State St., Iola, WI 54990-0001
Phone (715) 445-2214 • Fax (715) 445-4087

Credit Card Orders Toll-Free

(800) 258-0929 Dept. ZGB2

M-F 6:30 AM - 8 PM; SAT 8 AM - 2 PM, CT

Excellent Publications at Exceptional Values